BRITAIN
A GENETIC JOURNEY

ALISTAIR
MOFFAT

BIRLINN

first published in 2017 by
Birlinn Limited
West Newington House
10 Newington Road
Edinburgh
EH9 1QSs

www.birlinn.co.uk

ISBN: 978 1 78027 463 8

British Library Cataloguing-in-Publication Data
A catalogue record for this book is available from the British Library

Typeset by Iolaire Typesetting, Newtonmore
Printed and bound by Grafica Veneta

Alistair Moffat was born and bred in the Scottish Borders and studied at the universities of St Andrews, Edinburgh and London. A former Director of the Edinburgh Festival Fringe, Director of Programmes at Scottish Television and founder of the Borders Book Festival, he is also the author of a number of highly acclaimed books. From 2011 to 2014 he was Rector of the University of St Andrews.

Contents

List of Illustrations

✖

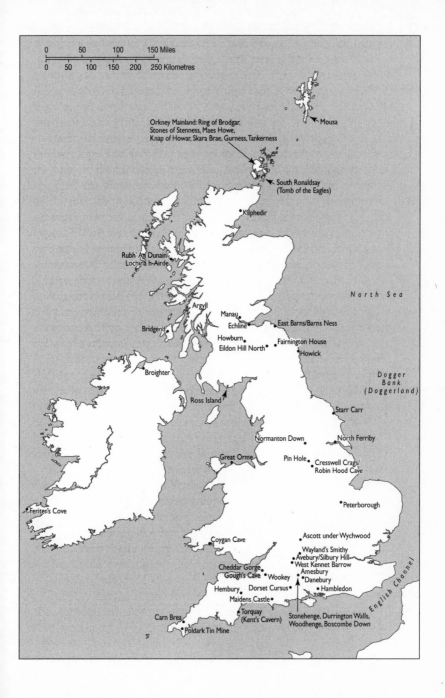

0   50   100   150 Miles

0   50   100   150   200   250 Kilometres

Orkney Mainland: Ring of Brodgar,
Stones of Stenness, Maes Howe,
Knap of Howar, Skara Brae, Gurness, Tankerness

Mousa

South Ronaldsay
(Tomb of the Eagles)

Kilphedir

Rubh' An Dunain/
Lochina h-Airde

North Sea

Argyll

Manau

Echline   East Barns/Barns Ness

Bridgend

Howburn   Fairnington House

Eildon Hill North   Howick

Dogger
Bank
(Doggerland)

Broighter

Ross Island

Starr Carr

Normanton Down   North Ferriby

Great Orme

Pin Hole   Cresswell Crags/
Robin Hood Cave

Feriter's Cove

Peterborough

Coygan Cave   Ascott under Wychwood

Wayland's Smithy
Avebury/Silbury Hill
West Kennet Barrow
Amesbury
Danebury

Cheddar Gorge

Gough's Cave   Wookey

Hembury   Hambledon

Maidens Castle

Carn Brea

Torquay
(Kent's Cavern)

Stonehenge, Durrington Walls,
Woodhenge, Boscombe Down

Poldark Tin Mine

English Channel

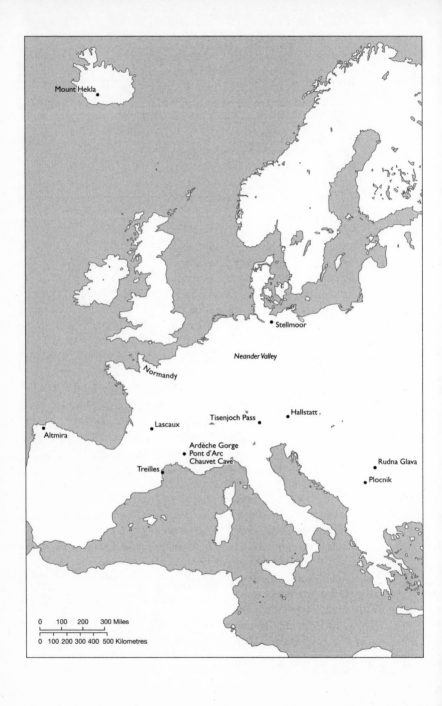

Mount Hekla

Stellmoor

Neander Valley

Normandy

Lascaux

Tisenjoch Pass

Hallstatt

Altmira

Ardèche Gorge
Pont d'Arc
Chauvet Cave

Treilles

Rudna Glava

Plocnik

0   100   200   300 Miles

0   100 200 300 400   500 Kilometres

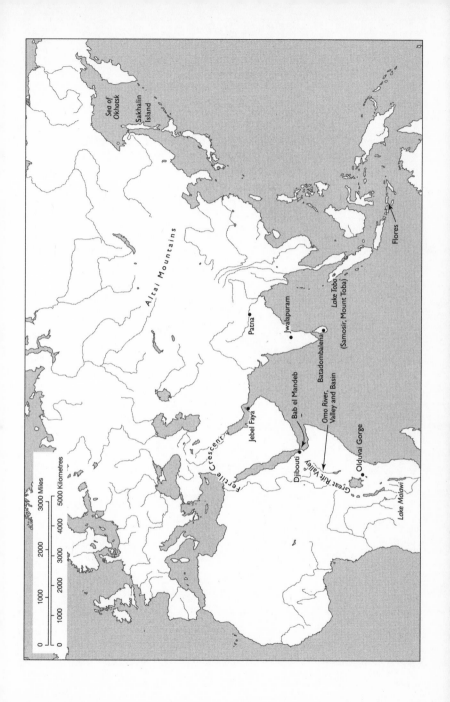

Preface

✖

WHEN HERODOTUS of Halicarnassus wrote history in the fifth century BC, he intended it to be something like an investigation, the collation of statements about events, people and circumstances made by those who had been there and seen them. Like a detective, he was pursuing enquiries. In Greek, 'history' meant something like 'testimony'. Soldiers who fought in battles, travellers who had seen Nile crocodiles, supplicants who had bowed before Persian kings – they were the sorts of people Herodotus wanted to hear from. But as timelines lengthened and perspectives shifted, historians inevitably came to depend on more distant sources, usually the written records of what witnesses or the actors in important episodes said about them. And for many centuries, most of the writing in Britain was done by clerics who were chroniclers often far removed from the action, rarely actual witnesses. Into the modern period archaeology came to supplement the patchy survival of documentary records, and for the long millennia before Herodotus and the statements of witnesses, what could be excavated and reconstructed became virtually the only source of reliable information. Without the patience, skill and imagination of generations of archaeologists, our prehistory would amount to little more than a set of assumptions and guesses.

Recently, population genetics and in particular the study of ancestral DNA have added an entirely new dimension to our understanding of our past. The ability of scientists to identify the origins and dates of DNA markers and to use them to track the movement of people across the Earth has been revealing, sometimes startling.

The dim and very distant prehistoric past can come brilliantly and movingly alive when the passage of a marker is traced from Manchuria and the shores of the Yellow Sea in the North Pacific clear across the Eurasian landmass to be found in Edinburgh in 2013.

For many millennia the last ice age held Britain and much of the northern hemisphere in its sterile, savage grip. In the brilliant white landscape of the ice-domes, frozen mountains a kilometre thick where incessant hurricanes whipped around their flanks, nothing and no one could survive. That brute fact made Britain a clean slate, a place waiting for its people and their DNA to come. When temperatures at last began to rise, the glaciers rumbled, groaned, splintered and cracked, and the land greened once more, people returned. These pioneers, the first of our species to see the rivers, the grasslands, the forest canopy and the hills and mountains that had been sleeping under the ice-sheets, are amongst our earliest ancestors. They may seem like another race, but DNA confirms an unbroken link. And when the last of the ice had gone, more people came north to hunt and gather in the wildwood, to fish and forage on the seashore, and to begin Britain.

Over thousands of years of prehistory, our ancestors walked to the farthest north-west of the Eurasian continent. DNA maps their movement and can approximately date their arrival. For at least four millennia after the ice, Britain was a peninsula and our people could make their journeys dry-shod. The landscape they entered was virgin. The ice had swept all life out of Britain and after *c* 9000 BC the land waited for the hesitant tread of the earliest pioneers. That simple fact makes our islands the sum of many migrations, a destination at the farthest north-west reach of Europe for many genetic journeys. A nation of immigrants on the edge of beyond.

All human beings, indeed all living organisms, have DNA, and that means something unarguable: we are all part of history. When Herodotus' witnesses reported, they tended to remember the doings of the great or the mighty, kings, queens, warriors, battles and politics. But DNA makes us all witnesses, everyone who lives in Britain. Neither spectators nor the crowd barely discernible in the dimly lit background, we are – every one of us – actors in the unfolding drama of our history.

The Irish Sea Glacier

One of the most striking images in studies of the last ice age is what some scientists believe should be called the Irish Sea Glacier. They believe that it rumbled for 700 kilometres from its source in the ice caps of Scotland and Ireland and then squeezed between higher glaciated land either side of the North Channel. It pressed on the northern coasts of Cornwall and the Scilly Isles. It is even conjectured that the glacier continued to flow south even when parts of south Wales, the Bristol Channel and the coasts of south-west England were ice-free. It was a kind of 'valley-glacier' of the sort seen in the Himalayas. Others disagree that such a glacier ever existed, but the value of such speculation is that it forces us to visualise how different familiar landscapes and coastlines were in the deep past.

'The past is a foreign country, they do things differently there', was the opening sentence of L.P. Hartley's *The Go-Between*, and it is apposite in considering the mysteries of our prehistory. Human sacrifice, cannibalism and the decapitation of children are all abhorrent practices now, but they were part of the experience of our direct ancestors, people whose DNA many of us carry. Fragments of their lives, like the unmade pieces of an archaeological jigsaw, lie around us on every side and as we try to make a clearer picture of the ranges and camps of hunter-gatherers, the great timber halls of the early farmers and the rituals that took place inside the stone circles, we must never fall into the trap of thinking of our ancestors as foreign. As Hartley wrote, the past is foreign, but its people were our people and their story is seamless, part of our story.

The balance of this book is heavily skewed to our prehistory for the excellent reason that the nine millennia before written record form the overwhelming proportion of our history. The early immigrations occurred before the brief visit of the Romans in the first century BC, and it is in tracing these that DNA studies can be most illuminating. But there is also a good deal to say about the first millennium AD. DNA can shine a light on what used to be known as the Dark Ages, the time of the Anglo-Saxons, the Vikings and other Northmen. After 1066, the picture stabilises as significantly fewer new people came and settled.

Since the industrial and agrarian revolutions, and the beginning of affordable sea and rail travel, Britain has seemed to become restless once more as many people moved. Across the Atlantic and further, from countryside to town and city and from upland to lowland, it appears that our people have left their native places in large numbers. But these changes seem more apparent than real. When those who have their ancestral DNA analysed are asked to name and locate their grandparents, it turns out that the demographics of Britain have shifted less than we think. When a Pictish marker was identified in 2013, many of those who carried it still lived in the ancient territories of Pictland in central and north-eastern Scotland.

More recently, Britain has seen an obvious influx of immigrants. After the Second World War, significant numbers of black and brown people have come mainly from the Caribbean and southern Asia to settle. They have not always been welcome and occasional outbursts talk of a threat to the British way of life. Not only is this shameful, it is inaccurate. At the outset of the twelfth millennium after the ice, Britain continues to be the destination of immigrants, just as it has been throughout all that long time. Incomers are not a threat to the British way of life, they *are* the British way of life. Our islands have been constantly enriched and renewed by the arrival of immigrants, they and their DNA have always added to the sum of what we are.

Part 1

1

Origins

On any London Monday morning packed trains rattle into King's Cross, Euston, Cannon Street and Waterloo. The brakes hiss, the doors open and Saracens, Saxons, Berbers, Cave Painters, Vikings, Angles and Picts pour out onto the platforms. On any Saturday afternoon at Ibrox, St James's Park, Old Trafford and Anfield crowds of Caledonians, Deer Hunters, Kurgans, Iberians, Rhinelanders and Anatolians roar on their teams, passionate in support, their sporting allegiances central to their identities. On any weekday morning all over Britain the school run delivers the children of the First Farmers, the Shell Collectors, the Foragers, the Shebans and the Yenesei to the gates of the playground.

These are the British, named by their DNA markers, all of them immigrants, all of them descendants of men and women from somewhere else, from the distant, shadowy millennia of deep time, the survivors of many epic journeys lost in the darkness of the past.

But now they are found, their stories lit by DNA, by the alchemical ability of geneticists to find traces of our history inside us, an immense story printed in the letters of our genome. DNA offers a new narrative, the unfolding – at last – of a people's history of Britain, a story of all of us who live on these islands at the end of Europe.

The story of the discovery of DNA itself is much younger and in some ways no less dramatic. In February 1953 two excited young men burst into the Eagle pub in Cambridge and announced to the lunchtime clientele that they had discovered the secret of life. If an eyebrow was raised perhaps it was because this was the sort of declaration young men make after spending a few hours in a pub. But in this case it was no less than the truth. Francis Crick and James D. Watson were researchers at the Cavendish Laboratory at the University of Cambridge, and that morning they had completed a model of the molecular structure of DNA, a model they knew was correct, convincing in every detail.

Deoxyribonucleic acid is indeed the secret of life because it is the basis of heredity, a biochemical blueprint for reproduction. The DNA molecule carries the patterns for constructing proteins, the building blocks of our bodies and the machines that run the cells that make up our organs. Every living organism has DNA, from bacteria such as anthrax to a whale, from the tiniest aphid to a giant redwood tree.

When Crick and Watson created a wholly coherent model of the molecule and comprehended how it copied itself, the clouds of conjecture cleared and new scientific horizons opened. Their discovery enabled the creation of entirely new academic disciplines such as the science of molecular biology. How hereditary diseases and disabilities are passed on was at last understood. And by understanding the DNA of diseases, effective means of combatting them could be found. Once they were able to recognise what proteins were deficient or missing, biochemists could manufacture drugs to deal with what had long seemed incurable. The likes of insulin for the treatment of diabetes was made possible by Crick and Watson's breakthrough.

When the two researchers pushed open the doors of the Eagle, they also wanted to celebrate a victory. They had won a race. Three universities had been competing to be the first to make what they all knew would be a momentous, world-changing discovery. At the California Institute of Technology the chemist Linus Pauling had adopted two approaches by developing techniques called X-ray crystallography and by building three-dimensional models. In 1951 he published his model of the protein, alpha helix. With all the resources at his disposal, it was surely only a matter of time before Pauling's techniques led him to a similarly credible model for the structure of DNA, and following the research at CalTech, it was likely to be a helix, a spiral curve.

4

X-Ray Crystallography

When Wilhelm Röntgen discovered X-rays in 1895, fellow scientists understood them as waves of electromagnetic radiation, another form of light. At the same time, it was recognised that crystals were regular and symmetrical arrangements of atoms and when X-rays were directed at them, the effects were seen to be very revealing about their atomic structure. A helpful analogy is the way in which the waves of the sea strike a lighthouse to produce secondary circular waves. In 1912 a German scientist, Max von Laue, shone a beam of X-rays through a copper sulphate crystal and recorded its diffraction on a photographic plate. This was a very significant advance in working out the atomic structure of matter, and in 1914 von Laue won the Nobel Prize for physics. At Cambridge University, William Lawrence Bragg and his father, William Henry Bragg, developed Bragg's Law, a method of connecting the scattering of X-rays with the structure of the planes within a crystal. It unlocked the atomic structure of molecules and minerals, and the earliest to be understood were those of table salt, copper and diamonds. The Braggs shared the Nobel Prize for physics in 1915, the only father and son ever to do so.

At King's College in the University of London, two brilliant scientists were collaborating, but not happily. Maurice Wilkins was a New Zealander who took a physics degree at Cambridge before the Second World War. Seen as a brilliant young scientist, he found himself working on the improvement of cathode-ray tubes for use in radar during the Battle of Britain. Later in the war Wilkins worked in the United States and was involved in the Manhattan Project. When it became clear that the primary aim was to build an immensely destructive atomic bomb, Wilkins (along with many other nuclear physicists) decided to turn his mind to other projects.

The supervisor of his PhD at Cambridge, John Randall, had also withdrawn from the Manhattan Project and together they began to work on X-ray crystallography, first at the University of St Andrews and later at King's College, London. Like Linus Pauling in California, Wilkins realised that these new techniques could be central to any understanding of the structure of DNA. There is a palpable sense of talented scientists turning away from the destruction of war to something more creative, optimistic and wholesome – the discovery of how life is made.

At King's Wilkins was joined by a remarkable woman. Rosalind Franklin had also contributed to the scientific effort behind the Allied victory in the more mundane field of investigating the properties of graphite, work that would eventually lead to the manufacture of carbon fibre. After 1945 she gained valuable experience in X-ray crystallography at the Laboratoire Centrale des Services Chimique de l'Etat in Paris before John Randall brought her back to London to work with Maurice Wilkins on DNA.

The collaboration was not a success. Franklin and Wilkins disliked each other so much that the pace of their research slowed and academic plotting and bickering sometimes seemed more important. Rosalind Franklin's response was to do her research alone, and from her notes it appears that she was closest to understanding the structure of DNA. Using diffraction techniques where a beam of X-rays is shone at a DNA crystal and the resulting reflections are captured as a series of dark or grey bands to produce an image, she successfully photographed the DNA molecule early in 1951. Analysis of the image clearly showed that it was a double helix, two spirals and not three as Francis Crick, James Watson and others believed it to be at that time. In her notes, Franklin wrote:

Conclusion: Big helix in several chains, phosphates on outside, phosphate-phosphate inter-helical bonds disrupted by water. Phosphate links available to proteins.

It was at this point embarrassment and academic pique came into play. Maurice Wilkins remembered James Watson sitting in a lecture given by Rosalind Franklin, 'he stared at her pop-eyed and wrote down nothing'. That turned out to be a crucial lapse. Two weeks later Crick and Watson announced that they had understood at last how DNA was structured, but omitted to mention that their solution was partly based on Franklin's lecture. The problem was that Watson had been forced to rely on his memory – and it turned out to be faulty. The model was structured around a value given by Franklin in her lecture and Watson had got it wrong. When she travelled up to Cambridge with her colleagues to see the new model, Wilkins recalled that Franklin 'all but laughed out loud'. It was mortifying. Professor Sir Lawrence Bragg, the Director of the Cavendish Laboratory, was not pleased and he instructed Crick and Watson to suspend their research. DNA was clearly something King's College understood far better.

This excruciating episode set a bitter tone. For a year Crick and Watson were frustrated, forced to work on haemoglobin (the protein that carries oxygen in our blood) rather than DNA. But then the instincts of competition came to their rescue. When Lawrence Bragg heard of the progress that Linus Pauling was making at CalTech, he changed his mind and asked Crick and Watson to revive their project. It was at this point human chemistry began to affect theoretical chemistry.

When James Watson went to London to meet Maurice Wilkins at King's College, he found himself arguing heatedly with Rosalind Franklin. Immediately after this altercation, Wilkins took Watson into another room where, without her knowledge, he showed him Franklin's latest and best images of DNA taken by X-ray crystallography. And around the same time more of her findings came into the possession of Crick and Watson. As a matter of routine, researchers at King's College wrote short abstracts on the progress of their work and Franklin's found its way quickly to Cambridge. It was not a private document and there was no suggestion of anything underhand but Franklin was apparently unaware that Crick and Watson had her research findings. 'Rosy, of course, did not directly give us her data. For that matter, no one at King's realised that it was in our hands', wrote Watson some years later. In 1961 Francis Crick admitted that 'the data we actually used' was the work of Rosalind Franklin.

Soon after Wilkins showed Watson the latest images at King's, Watson and Crick built their famous model. It resembled a twisted rope ladder with the uprights made from phosphates and sugars. But it included conclusions not visualised by Franklin. Francis Crick understood that the two helices, the spirals of DNA, twisted not in parallel but in opposite directions while Watson saw how the linked pairs of base compounds worked as the rungs of the rope ladder. This brilliant aperçu was the key to understanding how the molecule could copy itself, something even more critical than the structure itself. And when papers were published in the spring of 1953 in the leading academic journal *Nature*, only the three male scientists contributed and Rosemary Franklin was not acknowledged. In their famous, and very brief, article Watson's American exuberance about what he had realised about how DNA copied itself was tempered by the English reticence of Francis Crick and it is the subject of one of the greatest understatements in the history of science: 'It has not escaped our notice that the specific pairing we have postulated immediately suggests a possible copying mechanism for the genetic material.'

At first this earth-shattering discovery went entirely unreported. When Sir Lawrence Bragg announced at a conference in Belgium in April 1953 that his team had won the race to find the structure of DNA, there was no press coverage whatsoever. It was only when he spoke at Guy's Hospital in London a month later that Ritchie Calder, fortunately a scientist himself as well as a journalist, reported the breakthrough made in Cambridge in *The News Chronicle*.

As Crick and Watson, and to some extent Wilkins, received plaudits, Rosalind Franklin began to realise that she was seriously ill. Suffering from an aggressive form of ovarian cancer, she died in 1958, aged only 37. But surprisingly she appears to have held no grudges and when in a period of remission, she stayed with Francis Crick and his family in Cambridge.

A Nobel Prize can be awarded only to the living, and in 1962 James Watson, Francis Crick and Maurice Wilkins shared the award for medicine. Perhaps if she had lived long enough, Rosalind Franklin might have been similarly honoured for her visionary work.

In the decade following the discovery researchers began to understand much more about DNA. Crick and Watson's model showed how the DNA molecule unfurls, the rungs of the rope ladder separating down the middle into two half ladders, each with one rope and a half of every rung. Along with new phosphates and sugars to make the missing ropes of the ladders (as was mentioned in Franklin's note of 1951), new bases are then added to each half, and in this way two are created from one – the secret of life.

There are four chemical bases making up the rungs of the ladder; adenine (A), thymine (T), guanine (G), and cytosine (C). Each rung is made up of two sorts of base pairs. Adenine will only link with thymine and guanine only with cytosine. DNA from every living organism has these same ingredients but what makes us different from turnips or spiders is the order or sequence of bases. This is called the genetic code and scientists read it in the letters of the nucleotides, A, T, C and G.

For the decade following the creation of Crick and Watson's model, research concentrated on understanding how the body reads DNA to make proteins and how DNA is copied. The discipline of molecular biology developed quickly. But it was not until the 1960s that human population genetics began to evolve as a distinct academic discipline.

Born in Genoa in 1922 and having qualified as a doctor in 1944, Luca Cavalli-Sforza became interested in the new material on the analysis of blood groups in human populations that was published in the 1960s. It was not until the 1970s that the first sequences of DNA were read, but Cavalli-Sforza could see the value of using the classifications offered by blood groups. His aim was to use them to build evolutionary trees for *Homo sapiens* and to see how these linked and varied between different populations. But what puzzled him was the fact that blood groups were very much more diverse in Africa than they were over the whole of the rest of the world.

When scientists developed methods of reading the genetic code in the 1970s, it was at first a laborious and lengthy process and remained so until the early 1990s, but with the help of computing and other technological advances it accelerated and results began to appear quickly. Blood group classifications or markers were replaced by DNA markers. When all 6 billion letters of human DNA are copied in the act of reproduction, mistakes are occasionally made, and scientists noticed that some letters were out of place in a sequence. Because they are passed on to the next generation, these mistakes, these markers, allowed DNA to be used to trace lineages back into deep time.

Fred Sanger

Born in 1918 and raised as a Quaker, Fred Sanger was a conscientious objector during the Second World War, a time he spent not in prison but at Cambridge University where he began work that would benefit millions. Having completed his PhD in chemistry in 1943, he became fascinated by the problems of sequencing amino acids in bovine insulin. After a long string of spectacular success culminating in the award of a Nobel Prize in 1958, Sanger turned his mind to the sequencing of human DNA. With Alan Coulson and others, he ultimately came up with what is called the dideoxy chain termination method in 1977. A major breakthrough, it allowed long stretches of DNA to be sequenced quickly and precisely. In winning a second Nobel Prize (shared with two others) he joined Marie Curie, Linus Pauling and John Bardeen as the only double laureates in history. Having refused a knighthood because he disliked the idea of being called 'sir', Fred Sanger accepted the Order of Merit and allowed his name to be attached to the Sanger Institute near Cambridge, one of the world's largest genomic research centres. Without his pioneering work, DNA sequencing would still be laborious, time-consuming and disablingly expensive.

Meanwhile an answer to Cavalli-Sforza's conundrum about the diversity of African DNA was eventually supplied. It emerged close to Stanford University, where Cavalli-Sforza worked. At the University of California at Berkeley, a New Zealander of Scots extraction, Allan Wilson, was working on what he called the molecular clock. This was postulated as a means of dating the evolution of *Homo sapiens*, modern human beings, by looking at how DNA changed over time. Wilson and his team noticed that mitochondrial DNA, which is what women pass on to their children, mutated more readily and more regularly than the rest of our DNA. This made it easier to plot changes in mtDNA over relatively short periods of time, and not the millions of years of evolution conventionally envisaged.

This research led to a bombshell, and a solution to Cavalli-Sforza's puzzle. In the early 1980s Allan Wilson announced the existence of the woman he called Mitochondrial Eve, the mother-ancestor of us all. Using the molecular clock, he believed that it was possible to estimate the time and place where modern humans first evolved. About 150,000 years ago, Wilson asserted, all of us, from Apaches to Aboriginal Australians, from Scots to Zulus, descended from one woman who lived in east-central Africa. The announcement of the findings caused a sensation, and a very attractive Mitochondrial Eve and an Adam found themselves on the cover of *Newsweek* magazine.

PCR, LSD and Non-PC

Kary Mullis is an unconventional scientist. In 1993 he shared the Nobel Prize for chemistry for his improvement of a vitally important technique known as polymerase chain reaction. But in his Prize Lecture, he told his astonished audience that the award didn't make up for the fact that he had just broken up with his girlfriend. A year before he had started up a new business selling jewellery containing the amplified DNA of dead icons such as Elvis Presley and Marilyn Monroe. Mullis also reckoned that he would not have won his Nobel without the experience of taking LSD. Despite, or perhaps even because of these eccentricities, his improvements in PCR were epoch-changing. What Mullis was able to do was use an enzyme to bracket a DNA sequence and stimulate it to replicate almost an infinite number of times. This new technique had all sorts of applications and it allowed scientists to manipulate DNA to attack disease and to undertake complex research at much lower costs and achieve results quickly. It is a method used by all genotyping until very recently.

Wilson's theory ran aggressively counter to the conventional multi-regional view that *Homo sapiens* had evolved in different places from slightly different origins. In Europe it was thought that humans descended from Neanderthals, in China from Peking Man and in Indonesia from Java Man. But the new research insisted that we all have African ancestors, and a great deal of more recent work has supported Wilson's revolutionary view, although it is now recognised that a small proportion of the DNA of non-Africans descends from these other archaic humans.

Mitochondrial Eve is now thought to have lived approximately 190,000 years ago in east Africa, the area centred on modern Tanzania (although it must be added that evidence exists for a South African location for this prehistoric Garden of Eden, since the lineages of the Kalahari Bushmen and others are very ancient and very diverse). Fossil evidence confirmed the earliest appearance of modern humans, people who looked like us, at this time and as its techniques have developed, readings of DNA samples began to convert a theory into a fact. Researchers now believe that a man who might be called Y-chromosome Adam also lived in Africa, but not at the same time as Eve in a real version of the Garden. The ancestor of all men, traceable back through a Y-chromosome line, is thought to have lived some time around 140,000 BC probably in west Africa It is a misconception to believe that Mitochondrial Eve and Y-chromosomal Adam were the only men and women living at those times. Theirs are the only lineages that survive in the male and female lines, while others have died out. But it is, sadly, clear that Adam and Eve never knew each other.

Even Older

An ancient Y-chromosome lineage from Cameroon has been discovered in an African-American man from South Carolina and it matches that of four men from Cameroon. Labelled as A00, it appears to be very rare, but the startling finding is its date of origin. The oldest lineage was though to be A0 at *c* 140,000 BC, but researchers believe that A00 is much older, at *c* 237,000 BC. Work is ongoing.

As Cavalli-Sforza suspected from his study of blood groups, African DNA is much more diverse than anywhere else in the world, and many more markers are seen there. It seems certain now that the whole of the rest of the world was populated by men and women who walked out of Africa around 60,000 years ago. The probable reason for this ancient exodus is dramatic, emphatic.

In northern Sumatra the world's largest island within an island is green with lush vegetation, and the steeply pitched roofs of its Batak people punctuate the horizon. Samosir lies in the middle of Lake Toba, the biggest lake in south-east Asia at 100 kilometres long and 30 kilometres wide. The sharply pointed gables of the lakeside fishermen's houses mimic the prows of their boats and the brilliant greens of the scenery below are breathtakingly beautiful. But the beauty of the landscape is deceptive, for it is a memorial to a cataclysm.

Lake Toba shimmers quietly in the crater or caldera of a gigantic volcano. Some time around 73,000 BC, it suddenly exploded in a super-colossal eruption, an immensely destructive, climate-changing event, the largest anywhere on Earth in the last 25 million years. When Mount Toba blew itself apart, it may have obliterated life on our planet.

With a roar that must have been heard thousands of kilometres away, the volcano sent out 2,800 cubic kilometres of what geologists call 'ejecta'. Around 800 cubic kilometres of ash rocketed into the atmosphere to create a vast black cloud. High winds whipped up by the eruption quickly blew the ash to the west, out across the Indian Ocean. The year of this nuclear explosion may be only approximately dated but the season is certain. Toba exploded in the late summer, for only the monsoon rains could have deposited such a heavy and rapid fall of ash over the whole of southern Asia. A layer 15 centimetres thick has been calculated but at one site in central India archaeologists have recently found the suffocating grey blanket at 6 metres in depth. The ash covered vegetation of all kinds and the long nuclear winter that followed killed it.

High winds also carried and dropped huge tonnages of ash over the South China Sea, the Indian Ocean and the Arabian Sea. By screening out the sun and poisoning the water, the fallout from Toba killed plankton, sea vegetation, fish and larger creatures. Geologists believe that an even greater volume of volcanic ash may have fallen over the oceans than the land, but the effect was no less cataclysmic.

Around 10,000 million tonnes of sulphuric acid were thrown up into the atmosphere and some of it fell as black acid rain and devastated plants, animals and people. Pumice also shot high in the air and when it fell on the ocean, it instantly solidified into vast white rafts between five and ten kilometres across. These were picked up by the tsunamis that radiated from Sumatra and smashed into coastlines thousands of kilometres distant.

As thunder boomed and the Earth shuddered, red-hot lava spewed and poisoned rain fell, the eruption continued for two weeks. Sumatra was incinerated and covered by 2,000 square kilometres of boiling lava before the hollowed-out sides of the volcano collapsed in on themselves to form the caldera, what would much later become a beautiful lake. The fires caused by the eruption blazed over a wide swathe and sent vast plumes of smoke into the darkening skies.

As far away as Greenland, geologists have detected in the ice cores an abrupt change in the Earth's climate some time between 69,000 and 77,000 years ago. It can only have been caused by the destruction of Toba, and the cores show that what followed was indeed a long nuclear winter. A deadly sulphuric aerosol mixed with ash and smoke obscured the sun's rays and temperatures plummeted, particularly in the first three months after the eruption. What extended this half-lit, grey winter was the way in which the sun heated the aerosol, ash and smoke so that it rose into the stratosphere where no rain could fall to wash it out. This almost certainly caused a long period of nuclear darkness lasting perhaps ten or fifteen years. People must have thought the gods were angry and that the world was ending. If nothing could grow through the ash-covered ground, then animals and people could not hope to survive. Mount Toba may have almost ended the history of human beings, almost made us as extinct as the dinosaurs.

But the ash did not fall everywhere, and the dark blanketing of the stratosphere cannot have been complete – for human beings did survive. And Luca Cavalli-Sforza and Allan Wilson's research into African DNA suddenly appeared to connect with a recorded historical event. It seemed that the immense, world-wide destruction wreaked by the eruption of Toba was part of the reason why *Homo sapiens* and his (and her) origins are in Africa. It was the refuge where people survived the deadly fallout and the long nuclear winter.

Using computer models based on the number of markers seen in our genomes, geneticists believe that a tiny remnant, perhaps only 5,000 to 10,000 people, survived in the fertile rift valleys of east-central Africa. Other groups hung on in southern Africa and as far west as Cameroon. They can only have survived the horrors of Toba because the ash clouds and sulphuric aerosols did not obscure the sun completely and vegetation grew sufficiently for animals and people to carry on. Zoologists have noted that the East African chimpanzee, the cheetah and the tiger all saw their populations diminish drastically at this time, before they began slowly to recover.

As the Earth warmed and greened once more, the remnant groups across the continent also slowly recovered. They were hunter-gatherers who depended on a wild harvest of roots, fruits, berries, nuts and what animals they could trap or bring down, and because of their diet and way of life such communities could only grow very gradually. It may have taken many generations for there to be significant expansion, but after a time something surprising happened. A small group broke away from the east African communities and began to walk northwards. Perhaps only 300 to 500 people trekked out of the rift valleys. Geneticists are certain that the breakaway group was small because in their number only one mtDNA lineage that had descended from Mitochondrial Eve was present. All of the women in the world who are not Africans (and some who are Africans) are descended from this lineage, a marker labelled L3. And female descendants of L3, those in the two super-clusters of M and N, found across the whole of the rest of the world, are present in Africa now in only very low frequencies, and they appear to be recent arrivals.

A study led by Alon Keinan of Harvard Medical School suggests that more men than women walked northwards out of eastern Africa. By looking closely at variations in our X-chromosome and also at autosomal DNA, researchers have concluded that men were in a majority. No scientific reasons for this have been advanced beyond the sensible observation that in modern hunter-gatherer societies, women generally undertake short distance migration and men usually go on longer expeditions. It may be as simple as that.

After many more generations, the descendants of this small group reached the Horn of Africa, modern Djibouti. There the Red Sea narrows at the straits known as the Bab el Mandeb, the Gate of Tears. Now 15 kilometres wide and washed by treacherous rip-tides, it will have presented a much less formidable obstacle 60,000 years ago. Sea levels were lower then, the straits narrower and less deep, and there were small islands easily reached by rafts. A crossing could have been made in stages.

When our ancestors came ashore in the Arabian Peninsula, they stood on the edges of history. From these resourceful, curious, hardy and brave people the whole of the rest of the human race is descended. The DNA of all of us who are non-Africans was hidden in the genes of those who crossed the Gate of Tears and gained the farther shore.

2

Beyond Eden

As one of Europe's youngest nations and one with an ancient imperial past, Italy was anxious for status. Britain, France, Germany and even Belgium had divided almost all of Africa between them and, apart from tiny Liberia, the sole remaining independent state was the Christian empire of Ethiopia. In 1889 the Italians agreed the Treaty of Wichale with Emperor Menelik II, the King of Kings, the Lion of Judah, the descendant of Solomon. Unfortunately, it lost something in the translation. In the Italian language version, clause 11 was clear. It stated that Ethiopia would henceforth be a protectorate, represented by Italy in all its foreign relations. In Amharic, clause 11 was not translated literally. In fact it was different, much softer, more palatable to the Lion of Judah, and it merely advised that Ethiopia could use Italy to represent it in foreign affairs – if it chose to. The discrepancy soon came to light, erupted into war and humiliation for the new kingdom of Italy at the Battle of Adowa in 1896 when Menelik II led an Ethiopian army to a stunning victory over the Italians.

Far to the south an Italian soldier was to die in an altogether nobler cause. Captain Vittorio Bottego was the first European to see and survey the spectacular River Omo, and he and his expedition followed its course to the delta where it joined the waters of Lake Turkana. The Omo rises to the north, at 7,600 feet in the southern highlands of Ethiopia and quickly falls to 1,600 feet at the level of the lake. There are dramatic waterfalls, and in places the river raises long stretches of white water in its headlong rush to the south. Towards its delta, where the flow slows and the banks are low, the Omo can be very dangerous. It is the feeding ground of crocodiles.

The Queen of Sheba's Gold

Not only have geneticists discovered an mtDNA marker that arose in the biblical land of Sheba, archaeologists found the source of the queen's great wealth. Her realm straddled the Red Sea, taking in parts of modern Ethiopia and Yemen and when she came to Jerusalem and the court of King Solomon, she brought 120 talents of gold. Estimates vary, but a talent weighed about 25 kilograms or more. That is a great deal of gold and Louise Schofield has found the remains of an enormous ancient gold mine in the former land of Sheba. On the high Gheralta plateau in northern Ethiopia in 2012, she found a monumental slab or stele carved with a sun and a crescent moon, 'the calling card of the land of Sheba'. Then her team came across an inscription in Sabaean, the language the queen and her miners would have spoken. Known in Ethiopian tradition as Makeda, the queen ruled a civilisation that flourished for much of the first millennium BC and when she came to meet Solomon, she was said to have tested him with questions. Other accounts believe that the two monarchs became lovers and that the dynasty of Ethiopian emperors descends from their union.

Bottego was a cultured man, much interested in language and the peoples he encountered. He observed that the lower valley of the Omo lay at a crossroads in North-central Africa. At the northern end of the great Rift Valley, it was home to the Musi, Suri, Nyangatom, Dizi and Me'en peoples. All spoke different languages, and to Bottego's eye, they exhibited different characteristics.

Having arrived on the banks of the Omo on June 29th, 1896, the Italian explorer led his companions northwards a few months later, planning to travel through Ethiopia. But Bottego had no idea that a bitter battle had been fought at Adowa and that his country was at war. In the Maji Hills, not far from the Omo, the captain and his 80-strong expedition were ambushed by a much larger force of Oromo tribesmen. Along with several others, Bottego was killed. News of his death only reached Italy two years later through two surviving members of the expedition. They had been captured and imprisoned on the orders of Menelik II for what they later described as '98 days of terror'.

Safely back in Italy, Lamberto Vanutelli and Carlo Citerni were greeted as heroes, and even in the twenty-first century commemorative sculpture was still being erected with much pomp and ceremony. The two men 'narrated' *L'Omo: Viaggio di esplorazione nell'Africa Orientale*. Much of it is concerned with surveying, mapmaking, geography and politics; the southern limits of Italian influence needed to be mapped before it could be asserted. But Vanutelli and Citerni recalled something else, something that would prove to be much more durable than Italy's misplaced dreams of empire. The Omo Valley was like nothing they had seen before, at least in one remarkable aspect. The fossilised bones of human beings could be found there.

What revealed these deposits of our past was geology. Over aeons of time, many millions of years, the Great Rift Valley and the Omo Basin had convulsed with volcanic activity. Bulldozed by flowing lava, debris had been piled up all over the valley and all of this primeval fire and thunder created a layer cake of geological strata that, crucially, had been exposed by seismic faulting. And these layers, what would otherwise have remained buried deep in the ground, could be reliably dated.

K-Ar dating techniques measure the relative rate of radioactive decay of an isotope of potassium into argon and it is particularly accurate for periods far beyond the reach of other methods used by archaeologists, such as carbon dating. In the Omo Valley, where the layer cake had been helpfully sliced into sections by geological upheaval, the strata of volcanic tuff could be dated. And this in turn meant that the archaeology sandwiched between them, the remains of human fossils, could also be reliably dated.

In 1933 the discoveries of Vittorio Bottego and others were followed up by a French expedition led by Camille Arambourg. They mapped the area and recognised how important – and how extremely useful – the unusual geology of the Omo Valley was. During the Second World War the area was occupied by Allied forces who used the course of the river as a conduit to supply Ethiopian guerrilla units in their efforts to expel the Italian occupation.

Louis Leakey helped organise these supply lines. He was a

remarkable man. Born in 1903, the son of missionaries, his first instinct was to follow their calling. But he became fascinated with the life of the Kikuyu tribe in Kenya, made himself fluent in their language, had a hut built at the bottom of his parents' garden and went through a secret initiation ceremony. At the same time, this insatiably curious child became deeply interested in fossil-hunting, but before his passion could blossom further, Louis was sent to boarding school in England. From there he gained a scholarship to Cambridge University where he informed the authorities that his modern languages were Swahili and Kikuyu. By mistake he was invited to examine himself by conducting a viva voce. As an alternative, less partial means of validating his claims to fluency, Cambridge accepted an affidavit signed by a Kikuyu chief with a thumbprint. Having completed his studies, the young man longed to be back in Africa.

In 1931 Leakey led an expedition to the Olduvai Gorge in the eastern Serengeti Plains of what is now northern Tanzania. It was to become one of the most important prehistoric sites in the world, what newspapers would headline as 'The Cradle of Mankind'. The fossilised bones of animals had been found there but Leakey was sure that human remains could also be uncovered. Olduvai had been the scene of controversy. In 1913, on the eve of world war, the German archaeologist Professor Hans Reck had found a skeleton that he believed dated to 600,000 BC. There was uproar. According to convention, man had been created long after that impossibly early date and Reck began to think that he must have been wrong. In any event, war intervened and the colony of German East Africa became British under the terms of the Treaty of Versailles.

In November 1931, Leakey took Reck back to Olduvai and allowed him the honour of entering the gorge first. Within a day, as he had predicted to Reck in Berlin, Leakey found stone tools, evidence that early man had lived at Olduvai. These tools had sharp, skilfully worked edges, and sceptics began to see that they had been made by intelligent beings and were not geological accidents. Moreover, the tools had been found nine miles from where the stone had been quarried and this showed the clear ability of their makers to plan and organise their activities. It seemed that the Olduvai skeleton found by Reck was no freak and had been correctly dated.

After the Second World War, Leakey became involved in politics, steering a difficult course between his empathy for and understanding of Kikuyu culture and the rise of the Mau Mau movement in Kenya. He came to know Jomo Kenyatta, the leader of the Kenyan African Union, and did what he could to mediate during the state of emergency.

Against this unstable background, Leakey and his wife Mary spent as much time as they could at the Olduvai Gorge. In 1959 excavations began on a stratum called Bed 1, and very soon Mary discovered a fossilised human skull. It was nicknamed Zinj. A year later geophysicists K-Ar dated Bed 1 and found that it was 1.75 million years old. It was a sensation. No one had expected early man to have been on the Earth for such an immensely long time, and the discovery of Zinj, scientifically and securely dated, showed that Darwin's theory of evolution was no longer a theory.

The Leakeys became famous and their activities took place in a media circus with a constant stream of visitors. As more finds of fossils of early man were made Leakey's sons, Jonathan and Richard, became involved. After the granting of Kenyan independence in 1963, Jomo Kenyatta became Prime Minister, and three years later the Emperor of Ethiopia, Haile Selassie, arrived in Nairobi on a state visit. Louis Leakey was invited to lunch and the conversation turned to archaeology. The emperor enquired as to why there appeared to be no fossils in his country and Leakey assured him that there were. The problem had been the intransigence of Ethiopian bureaucracy when permissions to explore the Omo Valley had been sought. Haile Selassie made those problems vanish and in June 1967 the archaeologists began work.

Despite the territoriality and squabbling between the different elements of what was an international expedition (there were French, American and Kenyan parties), and the fact that crossing the Omo was made very dangerous by swarms of Nile crocodiles snapping at the boats (Louis Leakey counted 598 in one day), the archaeologists made hugely significant discoveries. However to Richard Leakey, put in charge of the Kenyan group by his father, these were disappointing. His excavators found two skulls but the dating of a clutch of oyster shells found just above them reckoned that these human beings had lived around 130,000 years ago. They were far too young to interest Richard Leakey, not the sort of sensational discovery made by his mother and father at the Olduvai Gorge. But in reality what he had found was just as important. It was the earliest evidence for the existence of our species, *Homo sapiens.* In 2004 the strata around the original finds was re-dated and the results pushed back the origins of our earliest ancestors to *c* 195,000 years ago. It seems that modern human beings first walked under African skies, and that the long journeys of our ancestors began there. And, crucially, this archaeological evidence chimed precisely with what Allan Wilson discovered in the late 1980s through his DNA research and an understanding of the molecular clock.

Preserved in the strata of the Omo Valley was evidence of the environment our ancestors lived in, a powerful sense of how fragile their existence could be. Plant and animal fossils suggested a changing climate with dramatic swings from an arid, sun-baked bush country after 185,000 BC, and before then, a wooded grassland that was home to antelope and other grazing herds. At other times there was a dense riverine forest on the banks of the Omo where wild pigs rootled and browsed. Researchers also believe that there were montane forests in the highlands where the river rises. The evolution of wild pigs could be seen in their fossils and the sequence helped date the swings of climate change and set a reliable chronology.

Lake Malawi is one of the largest and deepest lakes in the world, and 700 metres down in its black-dark depths lie more clues that can piece together a picture of prehistoric African climate. Cores drilled out from the layers of sediment on the bed supply a clear linear record of what life was like around the shores of the great lake. It appears that between 135,000 BC and 90,000 BC tropical Africa experienced severe drought. The levels of Lake Malawi dropped drastically by 500 metres and the landscape around it resembled desert. It is likely that to the north and south the Sahara and Kalahari deserts expanded enormously at this time. In such a hostile environment, one that lasted for 45,000 years, it is very likely that populations of *Homo sapiens* were very much reduced.

Our ancestors were hunter-gatherers and the stone tools found in the Omo Valley and elsewhere suggest that they existed on a wild harvest of fruits, roots, berries, eggs and whatever animals they could trap or hunt down. They probably lived in family bands, some as large as 40 or 50 individuals, but these groups grew very slowly. Infant teeth almost certainly found the wild diet difficult and breast-feeding continued far longer than it does in modern societies. It could be four or five years before infants were weaned, and during all that time nursing mothers remained infertile. That in turn meant a long birth interval, and given the much shorter lifespan of women in prehistory, a likelihood of only three or four children, not all of whom would have survived.

Amazonian Mash

Isolated kindreds deep in the jungles of Amazonia, people who were pre-Stone Age and used bows and arrows and blow-guns, have been documented as having larger families than the hunter-gatherers of Europe. Mothers had babies, toddlers and older children even though they were mobile and moved considerable distances in their wide ranges. Older children helped with their younger siblings and the food-gathering abili-ties of these peoples appeared to be unimpaired by having many children. They overcame the problem of soft infant teeth by boiling roots and making a mash by pounding them. Monkey brains were also given to toddlers for them to suck and mothers chewed tougher foods before spitting them into their babies' mouths.

Hunter-gatherer populations not only expanded extremely slowly, they were vulnerable, wholly dependent on the natural world for food, shelter, tools and firewood. The swings of climate change seen in the cores taken from the bed of Lake Malawi (and con-firmed by cores from Lake Tanganyika and Lake Bosumtwi) could cause populations to shrink dramatically or move long distances in search of better places to live. In the drought and desert conditions between 135,000 BC and 90,000 BC, there may have been only a few thousand of our species living in Africa.

At Jebel Faya in the United Arab Emirates, on the southern shores of the Persian Gulf, archaeologists have unearthed a puzzle. In an ancient rock shelter they discovered a set of stone tools: hand axes (flint stones with a sharpened edge but without a wooden shaft that could be held in one hand and used to chop), and an assortment of scrapers and burins or perforators. This toolkit resembled artefacts found in East Africa in the rift valleys, the Omo and elsewhere, but their date seemed much too early.

Using a technique known as luminescence dating, researchers from the University of London calculated that the axes and tools from Jebel Faya were knapped and used some time between 125,000 BC and 100,000 BC. Their manufacture, by people who could be described as *Homo sapiens*, predated the exodus from Africa in 60,000 BC by many millennia and their findspot lay more than a thousand miles from the Bab el Mandeb, the straits at the mouth of the Red Sea.

Hans-Peter Uerpmann of the Eberhard-Karls University in Germany led an investigation into the climate of the Arabian Peninsula *c* 130,000 BC. Not only did he and his team establish that sea levels were much lower then, allowing an easy passage across the straits, but they also found that the desert climate of modern Arabia was very different in the deep past. Significantly greater rainfall watered a much greener land and fed a landscape of lakes and river systems. And so it seemed that while central Africa grew dry, arid and desert-like, plants and animals flourished in the Arabian Peninsula. It was an environment where *Homo sapiens* could also flourish.

More archaeology on the shores of the Persian Gulf is suggestive of a better life beyond prehistoric Africa. Jeffrey Rose of the University of Birmingham and his colleagues have come across a series of settlements on the edge of the Gulf that date to *c* 7,500 BC. They turned out to be a great deal more sophisticated than expected, with well-built stone houses, evidence of long-distance trade networks, decorated pottery, domesticated animals and the remains of one of the oldest boats ever found. Rose believed that these early farming settlements lay on the edge of a huge oasis, a land watered by the rivers Tigris and Euphrates that is now submerged under the waves of the Persian Gulf. Palaeoclimatologists reckon that this fertile oasis was dry land *c* 100,000 BC and Rose's conjecture is that it may well have been home to a large population of *Homo sapiens* at that time.

Speleothems are mineral deposits found in caves. Perhaps the best-known varieties are stalactites and stalagmites. Formed by water seepage, they can act as a proxy record of climate change over long periods in the same way as ice cores. Researchers from the Hebrew University of Jerusalem have analysed speleothems from five caves in the Negev Desert and they are able to show that the weather was much wetter in that region between 140,000 BC and 110,000 BC. The Sinai-Negev landbridge between Africa and Near-Eastern Asia would therefore have been much more hospitable than the modern desert and would have allowed early movement of hunter-gatherer bands across it. At Skhul Cave in Carmel and Qafzeh Cave near Nazareth fossilised fragments of the skeletons of what appear to be *Homo sapiens* have been dated to *c* 100,000 BC. And it seems that all of these jigsaw pieces suggest that some of our ancestors moved north-east and east out of Africa long before Mount Toba blew itself apart *c* 73,000 BC.

However all that may be, the piecemeal evidence shows something unarguable: that human populations were fragile and very small, and more, that the present dominance of the Earth by our species was not inevitable. Indeed, climate change of the sort experienced by our remote ancestors may yet be our undoing.

When Toba erupted, it seems that these early pioneers on the Persian Gulf and what is now modern Israel did not survive the horrors of the nuclear winter that followed it. If there was an oasis in the Gulf, it lay close enough to the volcano to be badly affected and winds may have carried the deadly sulphuric aerosol mixed with ash and smoke north-westwards so that it darkened the lands of the Near East long enough to be fatal or provoke a trek back to Africa. As with much of our history, the movement of our species out of Africa may well be a story of advance and retreat.

After the cataclysm of Toba, the Earth's climate recovered, and so did other humanoid populations. Our species was not alone.

In 1857 it was believed that The Missing Link had been found. When Benjamin Disraeli described it as 'the link between apes and angels', he wasted no time in memorably placing himself 'on the side of the angels'. In the Neander Valley near Düsseldorf in Germany, quarrying displaced bones from a clay-filled cave in the steep sides of the ravine. Thought at first to be those of cave bears, they were retrieved and sent to Johann Carl Fuhlrott, a teacher and natural historian. He saw that the bones were in fact human and a year after they were found, descriptions of parts of what became known as Neanderthal Man were published.

Other skulls had been discovered earlier, in Belgium and in a quarry in Gibraltar. Labelled as 'ancient human' at the time, they were now recognised as Neanderthals. In the following 150 years, the bones of more than 400 Neanderthal men, women and children were identified over a wide range of Europe, the Middle East, the Caucasus and to the east of the Caspian Sea.

A picture gradually developed. Reconstructed skulls had massive brow-ridges and instead of the characteristic dome-shape of *Homo sapiens*, Neanderthals had lower foreheads and more elongated crania. Their chins receded and their noses were flat and large. All the supposed lineaments of gormlessness and a large part of the reason why 'Neanderthal' is now such a pejorative term. When a name for this separate species was being sought, a German professor, Ernst Haeckel, suggested *Homo stupidus*.

In fact the only persuasive reason for seeing Neanderthals as less intelligent than *Homo sapiens* is that they did not survive. It may be that our ancestors had a set of superior skills and understandings that simply out-competed them.

It is thought that the stocky, barrel-chested, bow-legged and short physique of Neanderthals was better adapted to cold climates, and that the large flat noses were efficient heat exchangers, able to reduce the shock of very cold air to the lungs. And it is certainly true that over most of their range, from Britain to the Caspian Sea and beyond, the weather was often much colder than it is now. Between 100,000 BC and 30,000 BC, the period into which almost all discovered Neanderthal remains fall, there were several long episodes when mean temperatures were very low. But over this immense stretch of territory, Neanderthals appeared to thrive and a total peak population of 70,000 has been estimated.

What would become Britain saw Neanderthal hunting parties roam the landscape. In 2002, the archaeologist and expert flint-knapper John Lord made several visits to a working gravel pit near the village of Lynford in Norfolk. As the diggers scarted out the gravel, John noticed what he was sure were mammoth bones and small flint hand-axes. Excavation was halted and in a hurriedly arranged rescue dig, archaeologists found the remains of nine mammoths and also woolly rhinoceros, reindeer, wild horse, bison, fox, wolf, hyena and brown bear. These were the discarded remnants of Neanderthal hunting forays, and it was once more recognised that a culture that could bring down such large and dangerous animals certainly possessed great skills and much courage.

In warmer episodes after 200,000 BC, Neanderthal folk came to Britain and evidence of occupation has been found in the Cresswell area, south-east of Sheffield at Pin Hole and Robin Hood caves. To the west, they also seem to have lived for a time at Wookey in Somerset and also at Coygan Cave in Carmarthenshire in Wales.

Neanderthal hunting methods have been conjectured from a careful examination of the evidence of injuries to their skeletons. They resembled closely those of modern rodeo riders, men who were often thrown from bucking, unbacked horses. From this key observation, a convincing hypothesis developed. The flat noses of Neanderthals may have been efficient heat exchangers but they were also capable of inhaling a much larger volume of oxygen. And their short, stocky, barrel-chested bodies, with tremendously strong arms, hands and legs, very much stronger than the limbs of modern humans, were adapted for a particularly brutal and dangerous way of hunting.

Neanderthals did not have bows and arrows and their stone- or flint-tipped spears had only a limited throwing range. That mean that they needed to get close to their prey, the huge hairy mammoths, the woolly rhinos, bison and wild horse, close enough to wound them badly. There then followed a bloody pursuit of a weakening animal that may have ended in a furious sprint. Neanderthal physique was well suited to short bursts of great speed and once they caught up with a fleeing, wounded animal, they jumped on its back to bring it down. Perhaps with their immensely strong hands and arms, they could force smaller prey, like horses or reindeer, to overbalance and fall. And once an animal was down, a successful kill would have been completed very quickly.

It is also believed that Neanderthal hunters laid ambushes, presumably on either side of tracks habitually used by prey animals or in places where migrating herds passed through and which offered good cover. It may be that in a large group, these men and women directly confronted those that were not flight animals, the likes of woolly rhinoceros and mammoths. Perhaps they fought them, jumping on their backs and jabbing at their heads and eyes with stone-tipped spears as the animals roared and trumpeted.

Neanderthal skulls suggest that their brain case was larger than that of *Homo sapiens* and their eye sockets were wider, implying bigger eyes and better all-round vision. Excavations across the European and Asian ranges have discovered several types of tools and it seems that they also buried their dead and could build shelters and rudimentary dwellings.

Our ancestors appear to have encountered Neanderthals face to face, or even more intimately, because DNA studies at the Max Planck Institute in Germany suggest that interbreeding took place. Between 1% and 4% of the human genome is thought to have come from Neanderthal DNA. This finding is disputed by some.

The Human Genome Project

Beginning in 1990 and led by Ari Patrinos of the US Department of Energy's Office of Science, this immensely ambitious project also included researchers and institutions from Britain, Japan, France, Germany and Spain. The goal was to sequence and map all 6 billion letters of the human genome and thereby enable advances in medicine and in science generally. Within ten years, President Bill Clinton and Prime Minister Tony Blair could announce the completion of a 'rough draft' of the genome, and all the sequencing was finally done by 2003. The results are stored in a database on the internet and they are available to anyone. The beneficial effects were immediate as, for example, companies created tests to show predispositions to illnesses such as breast cancer, cystic fibrosis and liver disease which in turn could be used to trigger early treatment and dramatically improve the chances of successful therapy.

The youngest skeletons of Neanderthals date to around 30,000 BC, the probable time of their extinction. Did our ancestors drive them out of their ranges, was there violent conflict or did disease destroy Neanderthal culture? While it is true that their physiques were well adapted to cold conditions, they may not have been able to cope well with extreme fluctuations in climate at the same time as sharing their ranges with incoming bands of *Homo sapiens*. If rapid climate change meant that familiar plants and the animals who browsed them were replaced or removed, perhaps within a generation, it may have been difficult for Neanderthal bands to find enough food. Bone analysis suggests that the Neanderthal folk of northern Europe were highly carnivorous (those of the Mediterranean ate fish, mussels, tortoise and rabbits). And if they were also being pushed to the margins of their ranges by the hunters of *Homo sapiens* groups, they may simply have been starved into extinction.

It is impossible to know exactly what happened, but it is certain that for thousands of years modern humans shared the planet with these fascinating people. And it is also certain that they were not stupid or somehow inferior beings. They were our ancestors, ancestors who lived alongside us for millennia.

High in the Altai Mountains of southern Siberia archaeologists worked in the damp and darkness of the Denisova Cave. It may have been an uncongenial place to spend many months, sifting through the levels of remains on the cavern floor lit by blinding arc lamps, but to a Neanderthal band and to our own ancestors, it was a good place to live. The interior of Denisova is large and it has a wide entrance easily reached across flat ground on the banks of a fast-flowing river, a reliable source of fresh water for the prehistoric occupants. And it was relatively watertight as it burrowed into the mountainside.

Having discovered that both Neanderthals and *Homo sapiens* had lived at Denisova at different times, the excavators from the Institute of Archaeology and Ethnology of Novosibirsk no doubt believed that their work might offer valuable insights into interactions between these two species of hominids. But nothing could have prepared them for what they found in 2008.

At level 11, dated to around 50,000 BC, the archaeologists carefully lifted out a fragment of fossilised bone. It was part of the fifth finger, the pinkie. Perhaps it had come from a member of a very early band of *Homo sapiens* that had occupied the cave as a hunting base or more permanent home. The Russians contacted the Max Planck Institute in Leipzig and sent a sliver of bone to see if DNA could be extracted. It could. The results were sensational, a discovery that only DNA could have made.

Geneticists were astonished to discover a new species of hominid, and one, like the Neanderthals, that had co-existed with the ancestors of modern humans. The fragment of bone came from a female who was immediately dubbed a Denisovan, a member of a different species whose existence had not even been suspected and would not have been discovered if its DNA had not been sequenced.

At the Max Planck Institute, scientists found that the mtDNA of the pinkie bone differed from that of modern humans by 385 base pairs (the base pairs being those that link the double helix of the DNA molecule) out of approximately 16,500 sequenced, while Neanderthal mtDNA differs from that of modern humans by around 202 base pairs. This implied that the Denisovans diverged about one million years ago. It also appeared that they were closer to Neanderthals than to *Homo sapiens*, and when this amazing discovery was made public, the Denisovans were thought of as east Asian cousins to the Neanderthals. And like them, they had certainly left Africa long before our ancestors.

The cool conditions in the cave had preserved the mtDNA in the pinkie bone to a surprising degree and in 2012 the German geneticists were able to do more detailed analysis and announce more surprises. The Denisovans had interbred with our ancestors, without question. Their mtDNA makes up between 4% and 6% of that of Melanesians, the native peoples of Papua New Guinea, Australian Aborigines and a scatter of small groups in south-east Asia.

This undoubtedly meant that the ranges of the Denisovans extended well beyond Siberia, far to the south of the Altai Mountains, over the Himalayan massif and down into tropical Asia and Australasia. It is certain, moreover, that when our ancestors walked out of Africa and moved eastwards in the millennia after the eruption of Mount Toba, they did not enter a virgin landscape. There were other footprints in the sand, other sorts of humans hunted the montane forests, lived in caves and fished in the rivers and lakes. Bands of Denisovans and Neanderthals are unlikely to have been large, but were wide-ranging over vast territories and where they walked and lived, left only gossamer traces of their passing. The discovery in the cave in the Altai Mountains could be seen as the archaeological equivalent of finding a needle in a haystack.

The first encounter between Denisovans and a band of *Homo sapiens* must have been astonishing – on both sides. Two groups of people who looked very like each other but who could probably not communicate easily, at least at first. Since they certainly had sex with each other, the ancestors of modern humans and Denisovans did eventually manage to develop some understandings, unless of course the sex was not consensual. And these encounters may have been brief and rare. To introduce between 4% and 6% of Denisovan DNA into the genome of Melanesian and Aborigine populations, it has been calculated that only 50 Denisovans needed to have sex, over time, with about 1,000 ancestors of modern humans.

The Max Planck Institute had done extensive sequencing and it produced more startling findings. From the ancient DNA extracted from the Denisovan pinkie, geneticists were able to say that the owner was a seven- to eight-year-old girl, had dark skin, brown eyes and brown hair. Reconstructions show that she did not look as different from modern humans as Neanderthals apparently did. Geneticists were also able to make comparisons between the DNA inherited by the young girl from her mother and her father. These suggested a population with very low levels of diversity and made it likely that the Denisovans developed from a small group of pioneers who had left Africa before expanding rapidly and widely.

The most intriguing evidence supplied by the extraordinarily accurate sequencing of Denisovan DNA sheds light on human evolution and hints at reasons why *Homo sapiens* out-competed all other forms of humanoid life on the planet. What geneticists call highly conserved genes, those vital to life and health, have been compared and they show important differences that may shed light on our recent evolution.

There are 23 highly conserved genetic regions in humans that differ from the Denisovans. Eight major differences have been detected. All relate to nerve growth and function in the areas of language and the ability to make connections, to link memories and

31

ideas in new combinations. Three of these that have undergone changes since the time of the Denisovans are related to autism and other language disorders. The geneticists at the Max Planck Institute believe that these genes that appear to have changed in the last 50,000 years may supply some explanation for the ability of modern humans to see a situation from the point of view of another person as well as the ability to conceal things and tell lies. In all of this remarkable analysis may lie the explanation of our survival, and the extinction of the Denisovans and the Neanderthals.

This pioneering study may also inform a debate. The beginnings of what might be called behavioural modernity in our species are much disputed. The new way of thinking and acting and the period when it came about have so far been characterised by the only evidence available, what archaeologists are able to uncover. And it seems that from approximately 50,000 BC onwards much more finely made stone and bone tools appear in the archaeological record as do hooks for fishing, evidence of long-distance exchange or barter for goods between different groups, the use of pigment such as red and yellow ochre for decoration, figurative art – both sculpture and painting – evidence of game-playing and music-making, food being cooked or seasoned rather than eaten raw and finally, evidence of human burial with some sort of ceremony or decoration, like the use of red ochre.

Proponents of the Great Leap Forward Theory argue that the emergence of this list is the result of an event, the consequence of major genetic mutation in our ancestors, the relatively rapid acquisition of different and new genetic faculties in the brain. Exactly the sort of thing discovered in the comparison between Denisovan and modern human DNA.

Others argue for process. Pointing out that there exists evidence for sophisticated artefacts before 50,000 BC, most notably in the Qafzeh Cave near Nazareth. There the remains of *Homo sapiens* were found, people who had left Africa much earlier than the ancestors of all non-Africans. The use of the cave has been dated to between 90,000 BC and 100,000 BC. There are also equally old sites in Africa that show at least some of the aspects of behavioural modernity. The Continuity Theorists argue that the sudden development of tool-making skills and the other attributes was not so much sudden as accelerated by population growth after 50,000 BC (and the gradual takeover of Neanderthal and Denisovan ranges) and a response to more difficult climatic conditions.

What is striking about these recent genetic findings is that they offer not only new evidence but something additional to the archaeology. After 50,000 BC, human brain function was certainly developing along different lines from that of other humanoids, something only DNA could reveal.

Big Boobs, Bellies and Bums

From a series of cave sites in southern Germany, small carvings of mammoth ivory have been found that date to the period before the last ice age. Some are of animals such as horses and mammoths but others, such as the Venus figurines, are of obviously female figures. They have large, exaggerated breasts, bellies and bottoms and their heads, arms and legs are often much smaller, almost an afterthought. And their vulvas are usually clearly explicit. In 2009 news was published of a much older female figure, a small ivory figure that dated to 35,000 BC. It may be the world's oldest depiction of a human figure. And with a similar set of characteristics as the Venus figurines, it is unmistakably a woman. Its function as a piece of art can be endlessly hypothesised but its date may be very significant. It appears to have been made around the time *Homo sapiens* was establishing communities in Europe and competing with Neanderthals. Was this figurine the earliest example yet found of our species' ability to think symbolically?

In 2003 an unexpected postscript was found. In a cave on the Indonesian island of Flores archaeologists uncovered the bones of what appeared to be another species of hominid. It became immediately clear that these people (the remains of nine were found) had been very small in stature, no more than a metre in height, much smaller than African pygmies. Their brains were also tiny, a quarter the size of those of *Homo sapiens*. And yet they did not appear to be primitive. Stone tools were found in the cave and the bones of a dwarf elephant known as a stegodon showed cut marks. There was also evidence of fire. Despite their small brains, it seemed that these people were capable of complex tasks and had developed co-operative hunting techniques.

Scientists labelled these extraordinary new finds as *Homo floresiensis* and the newspapers, of course, called them hobbits. Like the creations of J.R.R. Tolkien, they also had disproportionately big feet. Much debate ricocheted around the academic community but it is now generally accepted that the Flores hobbits were indeed another species of hominid. Astonishingly, they lasted longest, longer than the Neanderthals and the Denisovans, only becoming extinct around 10,000 BC.

Debate raged because, in part, it proved impossible to extract DNA from the hobbits. And in a remarkable incident, the Indonesian palaeoanthropologist Teuku Jacob removed the bones from Jakarta's National Research Centre of Archaeology and kept them for three months. During that time two leg bones were lost, skulls were badly damaged and a pelvis was smashed.

Scientists have been reduced to conjecture about the origins and nature of the hobbits. The most likely is that these isolated and remote communities evolved from a group of primitive hominids who left Africa around 2 million years ago, reaching south-east Asia and somehow surviving Mount Toba and all of the geological upheaval that preceded its eruption *c* 73,000 BC. And much later, despite the presence of *Homo sapiens* around them, the hobbits survived until very recently. Perhaps the fact that Flores was also home to a large population of giant rats, komodo dragons and even larger species of lizards deterred interference. It is thought that a volcanic eruption some time around 10,000 BC may have brought this extraordinary postscript to evolution to a final end.

But what is beyond conjecture is the fact that when our ancestors walked out of Africa *c* 60,000 BC, they shared the Earth with at least three other species of hominids.

Why did they leave? What prompted our ancestors to walk out of Africa? They were hunter-gatherer-fishers who depended absolutely on their environment to supply all of their food and when conditions were benign, it was probably a very congenial way of life. Compared to the constant day-in, day-out labour needed to sustain a society based on the seasonal rhythms of agriculture, the gathering of fruits, nuts, berries, roots and birds' eggs or the trapping of small animals or fishing took up less time and required less effort. When the weather smiled and the land was abundant, our early ancestors will have had time for leisure, time to talk, laugh and contemplate their world.

Around 50,000 years ago, long after the Earth had recovered from the devastation of Mount Toba, the archaeological record in Africa suggests that some hunter-gatherer-fisher bands were beginning to specialise. Tools such as harpoons or fish spears and hooks were made. Nets were woven, and it seems that some bands became more ambitious and went out after bigger game. It may be that a wish to reduce the amount of time spent finding food lay behind this shift, or perhaps populations were growing. In any event the urge to specialise is consistent with the general acquisition of behavioural modernity, the difference that showed up so clearly in the comparisons with Denisovan DNA.

Perhaps population pressure prompted movement. DNA can trace and date an early migration out of central Africa to the south. The first branch off the L marker of Mitochondrial Eve is concentrated amongst the San peoples of South Africa, who still live by hunting and gathering, and the Sandawe peoples of Tanzania in east Africa. It is L0, and also carried by the Mbuti pygmies of the Congo region, most of whom live in Zaire. They range in the Ituri, a tropical rainforest covering 70,000 square kilometres.

Other early branches of Mitochondrial Eve's L marker show a western migration to Cameroon and further, but L3, the mtDNA marker from which all non-Africans descend, appears to have moved northwards around 60,000 BC. These were the first steps in the epic journeys of our ancestors.

It may be that climate change also persuaded people to move out of central Africa. Hunter-gatherer-fishers are extremely sensitive to shifts in their environment, and if more or less rainfall or higher or lower mean temperatures altered the nature of the flora of a region, then the fauna quickly responded. If herds of prey animals were pulled north by climate change and the patterns of their seasonal migrations were re-set, then bands of hunter-gatherer-fishers would have followed. Sixty thousand years ago, people only moved for good reasons.

When our ancestors reached the Horn of Africa and the mouth of the Red Sea, they will have adapted. The seaside offered year-round supplies of food. Fish and shellfish were available and if, as they often did, bands chose to live near the mouth of a river (for fresh water), they could catch and enjoy both fresh- and saltwater species. Perhaps after a few generations of familiarisation with the tides and the seasons on the Red Sea coast, some curious individuals may have crossed to the Arabian shore.

The DNA evidence for such a crossing of the southern Red Sea is very persuasive. From the L3 marker, which originated in Africa between 60,000 and 70,000 years ago, the two mtDNA super-clusters of M and N are descended. Recent genetic research looked at people who carry old lineages that branch directly from the first non-African super-cluster of N. The results were emphatic and showed that there was indeed a very early lineage in the Arabian Peninsula that originated soon after 60,000 BC and that the pioneers had certainly crossed the Red Sea, probably at the southern end near the Bab el Mandeb, the Gate of Tears, where it joins the Indian Ocean.

Why did they cross? And how? Sea levels may have been much lower at that time and it may have been possible to get over in short hops, perhaps by poling or paddling rafts. The volcanic Hanish Archipelago lies north of the Gate of Tears and there the Red Sea widens considerably, but it may have been an easier crossing-point. Even in today's deeper waters there are a series of islets between the African shore and the main island. And its volcanic cone could have been kept in view at all points. Hanish itself is a large island able to support a substantial transient population.

The farther shore, the area that is now Yemen, must have looked like a promising destination. Otherwise why risk a crossing? Perhaps the pioneers sent scouting parties. It is difficult to imagine a migration on any scale and a potentially dangerous if brief voyage without knowing something of what lay on the other side.

NASA's Earth Observatory has offered some idea. Its photographs from space show clear traces of a landscape that was very different from the modern desert. The courses of many ancient rivers pattern the subcontinent, draining into the Persian Gulf, the Indian Ocean and the Red Sea. The beds of prehistoric lakes can be made out. Palaeoclimatologists believe that Arabia experienced a short-lived phase of wet climate around 55,000 BC. Perhaps those who climbed the barren slopes of the volcano on Hanish Island will have seen a lush and green horizon to the east.

Once across, our ancestors caught what historians, geneticists and anthropologists have all called the Coastal Express. It seems that within 10,000 years of coming ashore in Arabia, and perhaps sooner, bands of *Homo sapiens* were settling in Australasia. They advanced an average of 1 kilometre a year.

After what is known as the Single Southern Dispersal Event, populations must have increased rapidly. This may well have been a key factor in any competition with Neanderthals or Denisovans. By the time bands of people had moved along the coastal rim of Arabia and reached the Persian Gulf (which may have been a delta of the Tigris and Euphrates river system rather than a body of seawater), some carried on eastwards while others split off and travelled into what became the Fertile Crescent. Archaeological finds in the Indian subcontinent confirm the recent African origin of the migrants. Stone tools from digs at Patna in western India, Jwalapuram in south-east India and Batadomba lena in Sri Lanka are very similar in form and sophistication to those found in South Africa. The latter were discovered in the Blombos Cave on the Indian Ocean coast and at the Klasies River near the Cape. They were made by the southern migrants from central Africa, people who probably carried the mtDNA of the earliest branches from Mitochondrial Eve.

Archaeology and DNA analysis make a formidable inter-disciplinary combination in understanding how and when the world was populated by *Homo sapiens*, but there is another less certain but fascinating factor to take into account. Before it was recorded in writing, language left no material trace and yet its evolution can be interpreted so that it can say something about the speed and timetable of the Coastal Express.

Phonemic diversity is the key phenomenon. This measures the number of distinct units of sound (as opposed to written letters) in a language, the consonants, vowels and tones. And the modern pattern of phonemic diversity appears to mirror the routes taken by our ancestors after they walked out of Africa and crossed the Red Sea.

In the Kalahari Desert and in Tanzania, the Khoisan languages are spoken by the peoples enriched for the L0 marker, the first branch from Mitochondrial Eve. Known as click languages, they are the most phonemically diverse in the world. Taa is dying out, spoken now by only 6,000 people in Botswana and Namibia, but it can claim to be the most complex and varied language heard anywhere. Estimates vary, partly because those who have recorded it are not native speakers and there is also a considerable variety of forms even in such a tiny speech community. Nevertheless Taa makes the phonemic range of English seem impoverished. It has 141 phonemes, comprising 58 consonants, 31 vowels and 4 tones. These include between 20 and 43 click consonants.

The remarkable languages of the Khoisan speakers describe a hunter-gatherer life that has changed little over many millennia and it may even be heard as a faint echo of the speech of the first migrants, the carriers of the L0 marker, who moved south before 60,000 BC. Its diversity is a product of the range of vocabulary and nuance needed to describe and understand the natural world upon which these hunters depended completely. Flora, fauna, landscape and the seasons are all better described by those few languages that have remained in the mouths of rural populations such as some of the surviving Celtic languages of western Europe.

It may also be true that the onset of behavioural modernity and the development of complex language in Africa are coincident. In any case, just as human DNA is at its most diverse where *Homo sapiens* first emerged and has lived longest, so it seems that language is also extraordinarily diverse where it has been spoken longest – in Africa.

As bands of pioneers moved eastwards, they forced their way through a series of bottlenecks, episodes where, for a number of reasons, populations dwindled or even crashed. DNA can detect these where it finds a reduction in diversity, but so can language, particularly at the far end of the tracks of the Coastal Express. The Rotokas of East Papua use only 11 phonemes (written as 12 letters) and the Piraha people of the Amazon use only 10. Both of these languages are on the brink of extinction but have been extensively recorded. What they and others show is a substantially diminishing degree of phonemic diversity at the farthest extents of human settlement, a phenomenon that appears to walk in the footsteps of genetic diversity.

At 91 Ilsham Road, Torquay, in Devon, you will be greeted by Cavog. Sporting a thick black beard, spiky brown hair, wearing a tunic that may have been made out of animal hide and carrying a wooden club, Cavog is a Caveman. In a suburban street in Torquay, the entrance to Kent's Cavern can be found. Advertised as 'So Much More Than A Cave', it certainly is. In addition to being the site of woodland trails around the cavern, Devon cream teas, the Dragon's Lair and treasure hunts, Kent's Cavern was once home to the people who were amongst the first of our species to reach Britain.

Owned by the enterprising Powe family, the cavern is large and has been extensively paved with concrete, making it safe and accessible to many eager visitors. In 1903 Francis Powe used it as a workshop for making beach huts for use on Torquay's sea-front. But in 1927 a dramatic discovery was made. In a well-organised archaeological dig in the cavern, the Torquay Natural History Society found a fossilised fragment of an upper jawbone. It was thought to have belonged to a Neanderthal, and there was evidence that these folk had occupied the cavern. In 1989 the bone fragment was carbon-dated to between 34,400 BC and 32,700 BC. But in 2011 fossils from strata very close to where the jawbone was found were dated to considerably earlier, between 42,200 BC and 39,500 BC. From an analysis of the dental structure, the researchers also established something sensational. The fragment came not from a Neanderthal but from *Homo sapiens*. This made Kent's Cavern enormously significant: not only was it the first place in Britain known to have been home to modern human beings, it was also the earliest in north-western Europe.

This radical re-dating pulled the arrival of *Homo sapiens* in Europe back to 44,000 BC. It is generally assumed that bands of pioneers crossed at the narrowest point between Asia Minor and Europe, the Bosphorus. This seems unlikely. The Black Sea runs very strongly indeed through the straits, flowing into the Sea of Marmara before being narrowed once more into a torrent at the Dardanelles. And to make a crossing even more treacherous, there is also a mysterious undercurrent carrying water from the Mediterranean in the opposite direction, back into the Black Sea. The Bosphorus has long been recognised as a very dangerous stretch of water. In the third century BC Apollonius Rhodius recounted how Jason and the Argonauts had struggled to row through into the Black Sea:

> The narrow strait of the winding passage, hemmed in on both sides by rugged cliffs, while an eddying current from below was washing against the ship as she moved on.

Even mythical Greeks were probably better sailors than the pioneers of 44,000 BC, and it seems much more likely that a crossing from Asia Minor was made serially, by island-hopping around the northern rim of the Aegean Sea. This argues for a culture able to build and propel boats rather than using simple rafts, but of course, no material evidence of their seacraft survives. Once they came ashore in what is now northern Greece and were faced with the massif of the Balkans, bands of our ancestors may have coasted up the Black Sea and then moved into the heart of Europe by following the course of the Danube, probably not always on foot. Dated archaeological finds track this epic journey into history.

Neanderthal communities were already living in Europe at this time. There must have been encounters. In Aquitaine, in south-west France, there is a striking density of both Neanderthal and *Homo sapiens* sites, the highest recorded in Europe. And it seems that their population increased markedly after 38,000 BC, the time when the pioneers arrived. The number of caves or rock shelters rose from around 30 to 108 and the number of open sites from 7 to 39. It seems that in a relatively short period there was an explosion in population by a factor of 10, *Homo sapiens* hugely outnumbering Neanderthals in the region. Such an increase is thought to have been sustained by better hunting and gathering methods, more efficient food storage and processing (preservation by drying, smoking and roasting), enhanced mobility and better contacts within and between groups of *Homo sapiens*. The cumulative effect seems to have been inexorable and by 24,000 BC at the latest Neanderthals were extinct in Europe.

At around the same time, skies began to darken, storms blew

and the weather grew colder and colder. Winter snow did not melt in the spring but persisted all year round on high ground. Growing seasons shortened dramatically, vegetation shrivelled and as snow fell and the land began to freeze, the animals that depended on it retreated southwards. They were followed by the flight of the people who in turn depended on both.

Many scientists theorise that the weather grew quickly colder because the Earth had begun to wobble. Sometimes as it orbits the sun on its annual journey, our planet can wobble like a child's spinning top slowing down. Over time this tends to alter the angle of the Earth's tilt and can radically affect the amount and intensity of sunlight striking the northern hemisphere. A vicious cycle then begins. More snow accumulates, it reflects the sun's rays back out into space, insulates the ice it covers, encouraging it to thicken and spread south. The more snow and ice lay on the northern hemisphere, the more the area it covered advanced.

A huge ice sheet formed over Scandinavia, probably within only a few generations. It was two kilometres thick. Over the mountains of western Scotland an ice dome formed. Hemispherical and symmetrical in shape, its immense weight crushed the land below. As a focal point in the ice sheet that stretched from the coast of south Wales to north Yorkshire, the dome attracted most of the precipitation, rain, sleet and snow. Severe and constant winds were whipped up by the cold air around the summit and it flowed downwards to create areas of low pressure and dense cloud along the edges of the ice sheet. By comparison, up on the top of the dome it was almost always sunny. A brilliant white landscape of terrifying beauty.

The ice drove all life out of Britain. South of the diagonal line of the ice sheet, from north Yorkshire to south Wales, there stretched a polar desert where nothing and no one could live. For more than 10,000 years Britain lay empty, sleeping under the suffocating blanket of ice, waiting for her people to return. What follows is the story of their many journeys home.

3

Britain Begins

❈

WHEN THE LAST ice age came it may have come bewilderingly quickly, perhaps in the span of only a few generations. As the edges of the ice sheet and its polar desert crept ever further southwards, our ancestors fled, and almost all trace of their time in Britain was erased as the glaciers rumbled over the land. And as the cold gripped Europe, *Homo sapiens* sought refuge.

In the steep-sided river valleys on either side of the Pyrenees, people sheltered from the bitter winds and plummeting temperatures. More than 350 caves have been found in southern France and northern Spain, where bands of hunter-gatherers overwintered for many millennia. Known as the ice age refuges, they were home to the tiny populations of *Homo sapiens* that lived in western Europe 26,000 years ago.

Outside the caves storms blew and the world of ice and snow froze the landscape still. Winters lasted eight or nine months and temperatures rarely rose above freezing. The highest peaks of the Pyrenean ranges were encased in a brilliant white ice cap all year round, a constant reminder, visible on many horizons, that the climate could be even harsher. Down in the deep valleys, the refuges were sometimes shielded by dense stands of pinewoods whose branches will have bent low, loaded with snow. On windless days when the smallest sound echoed across the frozen rivers, the animals of the ice age searched the ground for food. Woolly rhinoceros, with its vicious horns, snuffled for the bitter grass. The huge wild cattle known as the aurochs, its seven-foot hornspreads swaying, browsed amongst the trees and on the valley floor, mammoths plodded on the gravel terraces, their long and dense hair brushing the lifeless ground.

Watching for weakness or older animals struggling to withstand the cold, predators patrolled the valleys. Cave lions, panthers, hyenas, wolves and bears waited. But another apex predator will rarely have ventured out in the extreme cold of the ice age winter. Instead, small bands of our ancestors lit fires at the mouths of their refuge caves, huddling for warmth, depending for most of their food on what they had dried and stored from the wild harvest of the summer to last through the long winter darkness. The constant demand for firewood meant that their ranges were wide and the lack of seasoned wood will no doubt have been a compelling reason for hunter-gatherer bands to move on.

As the snow slid off the pine branches and whumped onto the ground, and as the rivers began to flood with meltwater, the season all had been waiting for began. In a brief spring the world came alive again and it was the signal for much activity. Salmon and other species swam up-river to spawn and will have supplied easy and plentiful prey for those waiting on the banks or able to wade into the shock of the chill whitewater rapids. Migrating herds of reindeer, ibex, wild horses and other animals would soon move north for the sweet summer grass that was beginning to green the northern plains for a few short months. And when the approaching hoofbeats drummed and echoed in the steep-sided valleys, the hunters had to be ready.

The valleys not only offered shelter for our ancestors, they were also highways for the annual migrations of the prey animals they depended on. No doubt warned by scouts sent out to watch for the approach of the herds, hunters laid ambushes in the narrow defiles where animals were forced to go forward and could not scatter. Like the cave lions and the hyenas, they looked out for the weak, the young and the slow. Stragglers were no doubt especially vulnerable.

Those that waited in ambush probably did not yet have bows and arrows. Instead the hunters of the ice age refuges made strong stone- or bone-tipped spears. Fearsomely sharpened, the points were carefully carved and shaped so that the blunt ends were deeply notched to allow the wooden shaft to be secured more tightly with sinew thongs. The earliest throwing sticks, or atlatls, have been found in southern France and dated to around 19,000 BC; they were used to launch spears much greater distances than even the best throwing arm could achieve. The kill rate at ambushes in the valleys was probably not high, and animals not immediately downed would bleed to death as they tried desperately to escape. Those that survived the murderous gauntlet of hunters migrated to the great grasslands of the north. Treeless and with wide vistas, there these herds of flight animals were safe from all but the swiftest and most determined predators. But the hunters did not follow for they knew that when the summer forage died back in the autumn and temperatures began to drop, blood would run in the rivers once more as the onset of winter forced the herds to move south. And when the reindeer, deer and wild horses returned, they would have fawns and foals with them, small, leggy animals only a few months old and much easier prey for the spearmen.

It may be that the seasonal hunts involved more than near neighbours. The short summer was more than a time of plenty, it was a time of meeting, of finding marriage partners – and perhaps also it was a time of magic, music and ceremony deep in the caves.

The Ardèche Gorge in southern France is sometimes described and hyped in tourism blurbs as the Grand Canyon of Europe. And in truth it is spectacular – and it holds a secret that no blurb could over-hype. Thousand-foot high cliffs rise up straight above the river and nowhere is the geology more spectacular than at the Pont d'Arc. The largest natural bridge in Europe, an arch of limestone spans the deep river, channelling it briefly into a powerful, narrow flow.

On the afternoon of Sunday, 18 December 1994, three friends were walking near the Pont d'Arc on a path at the foot of the towering cliffs. Christian Hillaire, Eliette Brunel Deschamps and Jean-Marie Chauvet were amateur cavers, enthusiasts fascinated by the deep places of the Earth. And they were no dilettantes. For 30 years they had been looking for hidden caves in the Ardèche region, knowing that they walked in an ancient landscape, a place where several ice age refuges had been found.

A favoured means of detection was also delicate and difficult, and clues were easy to miss on windy days. As they walked beside the face of the cliff or even occasionally climbed it to reach likely locations, the cave hunters held the backs of their hands or their cheeks close to fissures in the rock. On their sensitive skin, they hoped to feel draughts of air from the deep, air that came from a hidden cave. And on that fateful winter Sunday afternoon, they made the discovery of a lifetime – of a generation – when they found what has been called the Cave of Forgotten Dreams.

When the three friends agreed that they definitely felt a powerful draught through a crack in the cliff face, they opened a narrow hole, barely large enough to wriggle through. Almost immediately they knew they had stumbled on something dramatic as their head-torches warned of a steep drop into the darkness of a wide and high chamber. The first hint that human beings had once occupied the cave was a wall of rock covered with what appeared to be small blocks of red pigment. On closer inspection, these were identifiable handprints pressed on to the surface in a clear and regular pattern. The highest could only have been made by a six foot tall man, according to later estimates by archaeologists. And they also noticed that he had a crooked little finger or pinkie.

As Hillaire, Brunel Deschamps and Chauvet moved deeper into the cave, their eyes opened wide with wonder. What they saw were some of the most stunning images ever made, and what turned out to be the earliest paintings in western European art. A panel of horse heads seen in profile, their manes flying, mouths open and whin-nying, galloped across one wall of the cave. Painted in a consistent style, this extraordinarily vibrant moment is believed to have been the work of a single artist, perhaps working quickly as the torchlight guttered, certainly inspired. On another wall two rhinoceroses fight, charging each other at speed, aiming their deadly horned snouts directly at each other, thundering across the imaginary landscape of the rockface, a scene that had been observed by the artist outside the cave. Lions, mouths agape and teeth bared, appear to race after their quarry and in another scene, a lioness seems to be refusing to mate with an amorous lion who rubs himself on her flanks. Bison, hyenas and other animals long extinct in Europe had been painted on the walls with extraordinary skill and brio.

What became known as Chauvet Cave, named after its finder, turned out to be large, almost 1,300 feet long and comprising two large chambers with connecting passages. The original entrance had been obscured by a rockfall some time around 24,000 BC, the height of the last ice age, and the cave had been sealed. Jean-Marie Chauvet and his friends knew that they had been the first people to look on the wonder of the animal paintings for 26 millennia. Like a time capsule, this cycle of images of the world of the early ice age, its animals and a pungent sense of its people, had been sealed, cast into darkness, hidden inside the great cliffs while the world outside changed utterly.

Despite the fact that archaeologists found traces of fires, they do not believe that people ever lived in Chauvet Cave. Instead, it seems to have been a black-dark temple buried deep in the Earth dedicated to the animals upon which the artists who painted them depended. Only visible when lit by torches, the bison, the cave lions and wild horses came alive in the flicker. Ceremonies may have involved music. Flutes made from the bones of large birds such as vultures and eagles dated to the same period have been found, and perhaps people whinnied with the galloping horses or bellowed and snorted with the fighting rhinos. Certainly the fires lit in the large chambers will have thrown the shadows of people on the cave walls and allowed them to run with or perhaps dominate the animals, or simply be part of the scenes.

But no one lived in Chauvet. Instead, the claw marks of cave bears have been found on the walls, some of them over the paintings. And nests on the clay floors have been recognised, places where the bears curled up and slept. Other marks of people are to be found, and handprints made by the six foot tall man with the crooked pinkie have been recognised in several places and at the farthest extent of the cave. Perhaps he was a painter-priest or a shaman.

It seems that Chauvet was painted in two distinct periods. The earlier was around 30,000 BC and the later around 25,000 BC. In the far chamber the archaeologists came across something haunting and moving. Walking into the darkness across the clay floor were the footprints of a little boy. He had been alone, carrying a torch, experiencing the magic of the cave and its animals, watching them animate as he moved forward, lighting them. The little boy, deduced to have been about eight years old, may have been one of the last people to see the paintings before the rockfall sealed the cave, allowing it to pass out of all memory for many millennia. Although the sequences of his paces show that he walked slowly, no doubt in awe of the marvellous images he saw on either side, there is evidence that he slipped on the damp floor. And at one point, his torch began to gutter and dim. Archaeologists have found fragments of charcoal where they boy scrubbed the dying embers against the cave wall so that they flamed up again.

He may not have been alone amongst all that wonder and mystery. Walking beside him are the clear pawprints of a dog. Not a wolf cub or a small wolf but what scientists now believe were the pawprints of a domesticated dog, a canid. Perhaps it was the little boy's companion in the enveloping darkness of the Cave of Forgotten Dreams. What is clear is that at between 20,000 and 30,000 years old, this trail is by far the oldest set of modern human footprints yet found. And even more remarkably, it may well be that the boy with the dog was the ancestor of many men living in Britain today.

During the lifetime of the communities that nurtured the little boy and for many millennia before, Europe's climate was very unstable. From analyses of the immensely long Greenland ice cores, it seems that between 45,000 BC and 12,000 BC cold and warm episodes alternated abruptly, with extreme oscillations taking place over short periods, sometimes no more than a decade. Icebergs were occasionally seen off the coasts of the Mediterranean. When the weather began to improve 14,700 years ago, it changed very quickly. Scientists studying the ice cores have detected a temperature rise of 22° Fahrenheit in the space of only 50 years, well within the reach of contemporary living memory. The effect must have been startling to those who had shivered in the ice storms that blew for 500 generations or more. And for the animals whose sensitivities to climate change were even more acute, it must have triggered migration. Over such a short period, their physiology simply could not adapt fast enough.

The first instinct of cold-adapted creatures such as reindeer, mammoth and others will have been to follow the cold as it retreated ever further northwards each year. And because these animals had been at the centre of the culture of the hunter-gatherers of the ice age refuges for so long, they were also pulled north to follow the hoofprints of the migrating herds. It may well be that the pace at which Europe was peopled when the ice shrank back after 12,700 BC was set not by people but by animals who migrated in search of the wide, open grassland pasture they felt safe in.

CO_2 Saved the World

Some time around 18,000 BC, the ice sheets that were spreading over northern Europe, the Americas and much of Asia halted. It was the beginning of the end of the last ice age. Palaeoclimatologists have looked at evidence from sediment cores drilled deep beneath the sea and in lakes as well as the tiny bubbles of ancient air trapped inside the cores. It seems that for reasons not yet explained the waters of the southern oceans began to release CO_2, enough to raise concentrations in the atmosphere by more than 100 parts per million over thousands of years. These levels are roughly equivalent to the rise of carbon dioxide in the atmosphere over the last two centuries. It seems that CO_2 was one of the reasons why much of our planet once more became habitable. Now it is seen as one of the gravest threats to our existence.

In another limestone gorge far to the north of the Ardèche, a landscape that would have been both familiar and attractive to the people of the ice age refuges in south-western France and northern Spain, the fascinating and grisly remains of some of our earliest ancestors have been found. In Cheddar Gorge in the Mendip Hills of the west of England, Richard Gough excavated a cave that would be named after him. Between 1892 and 1898, he made it safe enough to be opened to the public and even had electric light installed. Five years later, the oldest complete skeleton yet found in Britain was pieced together from bones found in the cave and, inevitably, immediately dubbed Cheddar Man. It is thought he lived some time around 7150 BC.

In 2010 more human bones were discovered in Gough's Cave and these were especially closely examined. They revealed a remarkable story. Much older than those of Cheddar Man, they dated to 12,700 BC. These were the remains of the pioneers, some of the first people to see Britain after its long 10,000-year slumber under the thick blanket of ice and polar desert. Brave and resourceful, no doubt, they established themselves in the vicinity of the limestone cave in Cheddar Gorge – where they were eaten by cannibals.

Using very sophisticated and innovative techniques to examine these ancient bones, scientists at the Natural History Museum in London have concluded that precisely the same techniques of butchery were practised on human beings as on animals killed by the hunter-gatherers. Barely discernible cut marks on the human bones showed that the same stone tools were used in both cases. The bodies of five people, a child aged between three and six, two adolescents, a young adult and an older adult, had been expertly defleshed, with all of the muscle and soft tissue stripped off the bones. And even the latter were cracked open or broken so that the nutritious marrow could be eaten. Brains appear to have been removed, tongues cut and eyes picked out. It seems that every part of the bodies of the child, the adolescents and the adults that could be eaten was eaten. Just as though they were the carcasses of reindeer or wild horses.

In view of all this meticulous butchery, it seemed unlikely to the team at the Natural History Museum that this very thorough form of cannibalism was carried out as part of a religious ritual, such as a form of ancestor-worship. While there may have been elements of ritual involved, and there is evidence that in the cave human crania were used as drinking vessels, it seems much more likely that human flesh was eaten in Cheddar Gorge because it was either relished as good and nourishing food, or because in time of great hardship and famine it was necessary to eat it in order to survive. Given that these pioneer bands were living and hunting right up against the ice and the polar desert, and in a climate that may have continued to fluctuate after the end of the last ice age, they may have been starving to death. Perhaps the child, the adolescents and the adults were themselves prey, prisoners who met a bloody end in the darkness of the cave.

Who were these people who butchered each other and ate

human flesh? They were probably our direct ancestors, people whose DNA we still carry, people from whom many millions of people in modern Britain are directly descended. A recent discovery in Gough's Cave underlines a cultural as well as a genetic continuity. In 2007 sharp-eyed archaeologists noticed an etching on one of the walls. It was clearly a mammoth, and its extravagantly long, curved tusks were particularly clear. No larger than a handprint, it was probably carved around 11,000 BC, just as the megafauna of the ice age were becoming extinct across Eurasia. Along with the mammoth, the woolly rhinoceros, cave hyenas, the Irish giant elk, the huge wild cattle known as the aurochs, cave lions and cave bears were all disappearing, unable to adapt to the warming climate. The etching of the mammoth in Gough's Cave would, like the paintings at Chauvet and elsewhere, have been vividly coloured with red and yellow ochre and other pigments. Its discovery shows a clear link between the pioneers in southern England and the people of the painted caves. And to modern sensibilities, a dichotomy. How could the recent descendants of the inspired artists who created the magical paintings at Chauvet and Lascaux also be cannibals?

DNA strengthens these connections. And mitochondrial DNA closely tracks the movement of people from the western ice age refuges to Britain. A very recent sample of 5,000 people from all parts of Britain, both men and women – since we all carry mtDNA – shows something remarkable. Just under 56% of all those tested in 2012 are matrilineally descended from those bands of hunter-gatherers who walked north across what is now France, or sailed up the Atlantic coastline and began to settle in Britain after the ice melted. Those who carry the DNA marker H and its sub-groups are by far the largest cohort at 44%, and markers U5 and V make up the remainder. The distribution is nationwide, from Orkney to Cornwall, and from East Anglia to west Wales. All of these markers appear to have arisen in the ice age refuges and then fanned out over Europe and Britain after *c* 9,000 BC. The highest modern frequency of H is in the Basque Country, at the western end of the Pyrenean ranges, a connection that makes a powerful ancestral link with the cave painters of Chauvet, Lascaux, Altamira and elsewhere.

It is important to bear in mind how very small hunter-gatherer populations were, numbering into only the hundreds over wide ranges and no more than a few thousand on either side of the Pyrenees at the height of the last ice age. But when the weather began at last to warm, the expansion over Europe was rapid and dramatic. The likely reason why such a tiny number of lineages, the mtDNA markers H, U and V, spread so far and so fast is not only linked to small populations, it is also a result of a demographic phenomenon known as the wave effect.

This was first measured in Canada. A huge study of the genealogy of around one million individuals in Quebec Province between 1686 and 1960 looked at the spatial dynamics of migration and their effect on the spread and frequency of DNA markers. Saguenay-Lac St Jean is about 200 miles due north of the city of Quebec. As waves of migrants began to move into the interior in the late seventeenth century, and accelerated in the nineteenth when the industrial felling of timber created many jobs, the DNA of those on the leading edge of the waves had a disproportionate impact, what is called gene surfing.

Probably through a process of rapid natural selection, those in the first wave appear to have been more fertile. It may be that the people who survived the tough, rough-and-ready life of the lumber trade were exceptionally hardy and that they passed on these characteristics directly to their many children. In any case their DNA markers multiplied at an appreciably faster rate than those they had left behind at the core, back in the city of Quebec. By tracing the genetic contributions of ancestors on the wave front and comparing them to those of the core amongst individuals born between 1931 and 1960, researchers found that those at the front had contributed genetically about twice as much to the current generation.

In Saguenay-Lac St Jean, female ancestors, those who passed on mtDNA to all of their children, had on average 15% more children and 20% more married children than those at the core. It is also significant that they were married about a year earlier than people in the city of Quebec. Over time these striking differences became more and more exaggerated, showing how gene surfing could distort the frequency of markers as they spread out from a core. It is very likely that a similar effect took place in the centuries after *c* 9000 BC as the pioneers repopulated a warming and greening Europe. Those lineages such as H, U5 and V that were carried at the front of the wave expanded enormously in frequency. That is why H and its sub-groups is found in 44% of modern Europeans.

Curiosity as well as necessity accelerated the speed of advance. People are and were insatiably attracted to the unknown, to discover what lay over the next horizon or around the far headland. As summers lengthened and grew warmer, expeditions will have set out. In the spring of 1993 archaeologists found a tantalising, tiny scrap of evidence for just such an expedition to the Hebridean island of Islay.

From cores pulled up from boreholes, geologists have been able to map the western edges of the great ice-sheet that crushed the north of Britain for millennia. It reached as far as the Rinns of Islay, what is now the westernmost peninsula of the island. Almost precisely on the edge of these ancient glaciers, a flint arrowhead was found. The best time for archaeological fieldwalkers to look for small objects such as coins, sherds of pottery or flints is after ploughing, and near the village of Bridgend, a group of them came across what is described by archaeologists as an Ahrensburgian point. It made a telling link and told a fascinating story.

The railway line that runs north-east from Hamburg in Germany to Lübeck passes a very narrow and steep-sided valley near the village of Ahrensburg. After the end of the last ice age, as southern Europe began slowly to warm, it formed part of the seasonal migration route to Scandinavia for vast herds of reindeer whose instincts led them north in search of a cooler climate. Lying in wait in this corridor were bands of hunters, and they held ready a weapon that was to revolutionise their way of life.

The remains of the earliest bows and arrows to be found in Europe have been uncovered at an excavation of their spring and autumn camp at Stellmoor, near the narrow valley, along with the bones of at least 650 reindeer. This speaks of organisation, of large numbers of hunters, of slaughter on an unprecedented scale. And the arrows that thudded into the shaggy flanks of the frightened, stampeding animals were tipped with razor-sharp, tanged Ahrensburgian points bound with sinew to pine shafts. These deadly points were very similar to the one picked up on that ploughed field on the Rinns of Islay.

Bows were found at Stellmoor but, sadly, destroyed in the bombing of Hamburg in the Second World War, well before carbon dating could have established how many thousands of years old they were. Other evidence points to the likelihood that the great slaughter at Stellmoor took place in the decades before the climate suddenly began to change once more, but it is impossible to be certain. What is much more clearly understood is that around 10,900 BC another ecological disaster was waiting to burst over the world.

Thousands of miles to the west of Islay, on the far side of the Atlantic Ocean, a vast lake of meltwater had collected across much of what is now the north of Canada. Known as Lake Agassiz, it held a volume of freshwater much greater than that of all the Great Lakes combined. And as the weather improved, the ice dams around it began to crumble. At first there was leakage to the south, sufficient to form the Mississippi Basin. But then, with a mighty roar, the dams to the north burst and within 36 hours the chill waters of Lake Agassiz flooded into the Mackenzie river system. Geographers from Sheffield University have traced what they call a mega-flood path. By looking at sudden cliff erosion in the Mackenzie delta over a range of altitudes, the only explanation that explained the evidence was indeed a huge flood, the emptying of the great meltwater lake.

This vast pulse of freshwater poured into the Arctic and through the Davis Strait into the North Atlantic. There it stopped the Gulf Stream dead, diverting it south and preventing its warming waters from reaching the British Isles and northern Europe. The effect was immediate and catastrophic. Mud cores taken from an ancient lake, Lough Monreagh, in County Clare in the west of Ireland, show that the climate cooled abruptly, over only a few months.

At a bewildering speed, the world changed utterly. As skies darkened, storms blew and snow fell, animals began to migrate southwards, driven by hunger. People followed and the north emptied once more. A huge ice-dome formed over Ben Lomond and much of Scotland, and the north of England became a pitiless white landscape, a desolate place where nothing, no animal or plant, could survive. It must have seemed to those who fled the icy grip of the intense cold that the world and its gods were angry.

For more than 1000 years, winter winds howled across the north and refugees sheltering in the south remembered their lost ranges. As these bands of hunter-gatherers retreated, they must have come into contact and perhaps conflict with the kindreds already established below the limits of the ice sheets and the polar tundra. Pressure to move back north may have been present for generations. And there are gossamer wisps of evidence that the early peoples of Britain may have attempted to venture back to their old ranges.

Norwegian scientists have gathered data that suggests what they call climate flickering, a phenomenon of the last few centuries of glaciation in the north. From two sediment cores pulled up from Lake Krakanes in western Norway and from the narrow continental shelf off the nearby North Atlantic coast, episodes of extreme variations in temperature have been detected. The lake core shows a random alternating pattern of increasing glaciation followed by rapid melting. And from the ocean core, a series of advances and retreats of sea-ice caused by the occasional resumption of warm and salty water flowing into the North Atlantic from the south.

The last may be seen as the faltering reappearance of the Gulf Stream. When Lake Agassiz burst through its ice-dams and a vast volume of very cold freshwater raced like a tsunami through the Davis Strait, it not only headed off the conveyor of warmer water from the Caribbean, it also diluted the salinity of the ocean. Between 70,000 and 150,000 cubic kilometres of meltwater made the North Atlantic significantly less salty, and this drastic change acted like a barrier for at least 1,000 years, maintaining polar conditions. When the natural processes of evaporation eventually rebalanced salinity, the Gulf Stream broke through and brought warmth once more to the north.

But the breakthrough may have been a stuttering, hesitant process rather than a single event. The evidence from the Norwegian sediment cores suggests decades when the weather improved, alternating with decades when the Gulf Stream was shut down and the ice came back and storms blew once more. Prehistoric hunter-gatherers mostly lived much shorter lives, with few reaching beyond 30 or more, and this will have meant that a generation could have at last moved north, settled and had children before being driven south once more. Implanted in the memories of children born in the north, will have been a lost land. Perhaps some adapted and on the southern fringes of the ice sheets and the polar deserts, they survived.

Even though temperatures plummeted by an average of 59° Fahrenheit (15° Celsius) in a generation in the period some historians have called the Cold Snap, Britain was not wholly abandoned at that time. Before the ice returned, hunter-gatherer bands had penetrated as far north as the Scottish Borders. Faint traces of what might have been a summer camp have been found at Howburn near Biggar. And 30 miles to the east of it, at Fairnington House near the River Tweed, a distinctive flint point has been found. This time it made a link not with the slaughter of reindeer in northern Germany but with an important early British site.

Yet another limestone gorge, Cresswell Crags, lies on the borders of Nottinghamshire and Derbyshire, not far south of Sheffield. Caves pierce the walls of the gorge and they gave shelter to nomadic groups of hunter-gatherers on summer expeditions from the south. Dated finds suggest that the Cresswell caves were occupied intermittently throughout the Cold Snap, from 10,800 BC to 9600 BC.

Many flint tools and animal remains have been uncovered at the Crags and the particular style of the former prompted Dorothy Garrod to posit a Cresswellian culture. This pioneering academic, the first woman to be appointed to a professorship at either Cambridge or Oxford, had, in the early 1920s, written a definitive treatment of the early repopulation after the Cold Snap, *The Upper Palaeolithic in Britain*. It stood as the standard text for a generation. What Garrod noticed was a distinctive style of knapping flints to make points and blades, a cultural habit that connected Gough's Cave in the Cheddar Gorge and Kent's Cavern in Torquay with the Cresswell Crags.

It seems that the first bands of hunter-gatherers were much further travelled than is often assumed. The quarries where the flints originated can be recognised and some of the tools discovered at Gough's Cave in Somerset came from the Vale of Pewsey in Wiltshire, a distance of almost 100 miles. At Cresswell, fragments of Baltic amber were found and although it could have passed through several pairs of hands before it reached the middle of England, there are other clear links with the North Sea coast. More amber and a scatter of non-local seashells suggest a very mobile prehistoric population.

Dorothy Garrod

Appointed as Disney Professor of Archaeology at Cambridge University in 1939, Dorothy Garrod found herself in an important, prestigious and surprising position. She had been awarded a chair eight years before women were allowed to take a full part in the government of the university. But she had gained her place on merit. Despite being shy and having to survive the vicious in-fighting that can disfigure university life, she had achieved a great deal since she entered Newnham College in 1913 to read for a Diploma in Archaeology. Degrees were not awarded to women until 1926. With the encouragement of Abbé Breuil, a pioneering (and devout) French archaeologist, she had discovered the Neanderthal skull called Gibraltar 2 in 1925. At Mount Carmel in Palestine, with an all-female team recruited from local villages, she worked out an immensely long sequence of intermittent occupation in a series of caves, contributing enormously to an understanding of prehistoric evolution. In her book, *The Upper Palaeolithic Age in Britain*, she coined the term 'Cresswellian' for the type of flints discovered in the caves near Sheffield.

Flint is heavy but nevertheless it seems that partially worked cores were carried from their source to be knapped as they were needed. But did our ancestors make all of these journeys, carrying loads, on foot? Needles hint at an answer. Several have been found at Cresswell and in their quiet way, they were revolutionary objects. Made from bone with eyelets wide enough to allow animal sinew to be threaded through, needles made life possible in the centuries after the ice. Between *c* 9600 BC and 8800 BC, Britain passed through a period known as the Pre-Boreal when the weather was dry but very cold. The ability to sew made close-fitting and warm garments possible for our ancestors, clothing that would not only keep out the wind and the chill but also not hinder the movement of hunters when they were out after prey.

Needles almost certainly also reduced the need to walk everywhere. Early boats were built by first lashing together a wooden frame of green rods (so that they retained a whippy tension) and then stretching over it a sewn membrane of animal hide before caulking the seams with pine resin or fat. No archaeological evidence of prehistoric hide boats has ever been found. They were constructed out of entirely organic and recyclable materials, as well as being light and insubstantial. But at Broighter on the shores of Lough Foyle in the north-west of Ireland, a farmer turned up a hoard of gold objects in 1896. One was a model of a sea-going curragh, and although it dates much later, to the end of the first millennium BC, the Broighter boat is sophisticated, the acme of a long tradition. An early Greek and later Roman commentators had much to say about the peculiarly British tradition of making boats from hide.

A scaled-up version of the gold curragh would have measured 15.5 metres in length and to make the frame rigid, it had as thwarts nine benches for eighteen oarsmen and a sail fixed amidships. But the earliest hide boats would have been much simpler. A smaller round version known as a coracle was, and still is, used for inland waters such as lakes, rivers and estuaries. It is likely that the far-travelled hunters of Cresswell, the Cheddar Gorge and Kent's Cavern used them to transport people and goods over long distances. When these expeditions ran out of water or encountered rapids or waterfalls, it was an easy lift to pick up a coracle and carry it. Upturned, they even made serviceable shelters. Whatever the realities, there is no doubt that our earliest ancestors ventured far in search of tools, prey and contact. And the use of boats would not only have made travel easier and faster, it may also help to explain the extraordinary speed of the advance north after *c* 9600 BC and the end of the Cold Snap.

Cresswell Crags, and especially the cave known as Church Hole, had another secret to give up. In April 2007 archaeologists began to recognise engravings of animals on the cave walls. Dating from *c* 10,800 BC, the onset of the Cold Snap, these included representations of bison and deer and what seemed at first to be long-necked birds, swans perhaps. Like the cave painters at Lascaux, Chauvet and Altamira, the Cresswell carvers used the natural irregularities in the surface of the rock walls to give seeming substance to their engravings. They were probably coloured with ochre but unlike the art in the darkness of the southern refuges, these images would have been lit by early morning sunshine.

Some who worked at Cresswell believe that the long-necked birds are in fact nothing of the kind. Instead, they interpret the engravings as human, or at least anthropomorphic, the outlines of naked women in profile, dancing with their arms raised up and their backsides thrust out. Perhaps life in the caves was not all about the gritty business of survival and the grim realities of finding enough food in the cold and dry Pre-Boreal climate. Perhaps there were some warm days, warm enough for naked dancing girls.

The speed of repopulation as the genetic wave surged northwards is made more than plausible conjecture by a startling image at Cresswell. The ibex is a species of wild goat with extravagantly long horns, and on the walls of Church Hole an artist has carved its outline. He or she must have working from memory, for the ibex was not native to Britain in the tenth and ninth millennia BC. Its habitat at that time was much further south in Europe. Near-contemporary remains of ibex have been excavated in Belgium and Germany.

A bone engraved with a superb image of a wild horse was an early Cresswell discovery, found in 1876 in the Robin Hood Cave. And when it, the mammoth engraving in Gough's Cave and one of a reindeer recently discovered in a cave in the Gower Peninsula in south Wales are all added to the images made at Cresswell, a credible cultural connection can be made with the cave painters of the southern ice age refuges. Once again, DNA reinforces what might be seen as tentative and makes another link that establishes the artists as our ancestors.

Two Y-chromosome markers still emphatically present in the modern population of Britain recall the journeys of the men who overwintered the age of ice in the caves on either side of the Pyrenees, the artists at Chauvet and those who came to Cresswell. Slightly more than 500,000 men, most of them Englishmen, are the direct descendants of many of those who moved north from the painted caves after *c* 9600 BC. About half, around 250,000, carry I-M284, a lineage that is common in south-western France, while the remaining half carry I-M26, a marker that is now massively present in Sardinia at a frequency of 37% of all men on the island. Both arose during the ice age and were probably carried by significant numbers of men in these tiny populations. The bias of M26 was possibly south of the Pyrenees, amongst the refuges of northern Spain, because when the weather warmed, people who carried it moved east around the Gulf of Lion and crossed the sea to settle in Sardinia. But some crossed or sailed around the Pyrenean ranges and eventually made landfall in Britain.

Ice Age Art

The railway engineer Peccadeau de l'Isle was working outside Toulouse on track construction when he decided to take some time off to indulge his hobby of archaeology. At the foot of a cliff near Montastruc, he found a stunning piece of prehistoric art, a sculpture showing two reindeer swimming. Carved from mammoth ivory, the end of a tusk, it represents two deer, one immediately behind the other, crossing a river. One is female, the other male and it appears that they were migrating because the sculptor has given the female a winter coat. Dated to around 11,000 BC, this beautiful object was made towards the end of the last ice age. Reindeer were vital to the survival of those who lived in the caves of the ice age refuges, and the Montastruc sculpture is not only beautifully made, it is also the sum of close observation.

Others walked, but not on water. As Captain Pilgrim Lockwood and the crew of the SS *Colinda* made ready at the south pier of Lowestoft harbour on a September evening in 1931, they could have had no inkling that they would retrace the footsteps of some of the first people to come to Britain, some of our very earliest ancestors. Once Lockwood had cast off and nosed his boat between the twin lights of the foghorn pierheads, he set a course for the Leman and Ower Banks. Lying about 25 miles off the Norfolk coast, this was a reliable fishing ground; the sea was shallow, and the seabed flat and unlikely to snag the wide net of the trawler as it scarted through the dark waters of the North Sea. Until that night, Lockwood and his crew were entirely unaware that less than 20 metres below their keel lay a lost landscape, the sunken geography of a forgotten subcontinent.

When the trawler cut its engine to slow down and begin to wind in the net so that it could be hoisted over the deck, the crew noticed a large, dark lump of what they called moorlog amongst the silver glint of wriggling fish. Clods of what looked like peat were sometimes dragged up from the sea bed and then heaved back over the side. But this was a big piece and Pilgrim Lockwood's log noted what happened next as he tried to break up the moorlog:

> We were halfway between the two north buoys in mid-channel between the Leman and Ower . . . I heard the shovel strike something. I thought it was steel. I bent down and took it below. It lay in the middle of the block which was about 4 feet square and 3 feet deep. I wiped it clean and saw it was an object quite black.

Once it had been thoroughly scoured, Lockwood's hard, black object turned out to be an ivory-coloured spear-point, what would be immediately and mistakenly described as a harpoon. Made not from steel but from the antler bone of red deer, the point measured more than 21 cm (8½ inches) in length. Along one edge a skilful craftsman had incised a series of barbed notches, probably cut with razor-sharp flints, and towards what was the bottom end of the blade, he or she had grooved several rows of transverse lines. These last supplied a clue to the precise use of the carved antler bone. It almost certainly formed the business end of a leister, a fish spear, and the grooved lines helped a hunter securely haft the point to a wooden shaft. Much later, the point was carbon-dated as early as *c* 9000 BC.

But what was such an ancient object doing embedded in the middle of a large clod of moorlog 25 miles from land, on the bed of the North Sea? In fact, how could it be that peat, compacted, decayed vegetation, was occasionally dredged up by fishing boats? The answer tracks back to the ice age and its after-effects.

So great had been the weight of the huge ice-domes over Scandinavia and the Shetland Islands, that they had pressed down hard on the crust of the Earth. This produced a geological phenomenon known as a fore-bulge, forcing the land to the south of the domes to rise up. Rather like the effect of a fat man sitting on one end of a bolster, making the rest of it bulge upwards. This was further exaggerated by the fact that so much of the world's water had been locked up in the ice that in the millennia after it began to melt, the levels of the oceans were much lower, about 120 metres lower than in the twenty-first century.

This meant that the vast area of what is now the North Sea was dry land. Between *c* 9600 BC and *c* 5000 BC, it was possible to walk from the European continent to Britain. And not by crossing a narrow landbridge or isthmus, but a huge landmass that at its greatest extent was much larger than Britain and Ireland put together. The long history of this lost land is a vital but mysterious part of the prehistory of Britain. It may have been transitory but it was not a transit area. This subcontinent must have been home to its own indigenous cultures, a place where more than 200 generations lived and died, somewhere our earliest ancestors hunted, gathered wild food and raised their families. After the Dogger Bank, a 60-mile-long sandbank in the middle of the southern basin of the sea that was once a range of hills, researchers have called this Atlantis to the east Doggerland.

After Pilgrim Lockwood clashed his shovel through the clod of moorlog, other tell-tale objects were found on the bed of the North Sea. The bones of mammoths and lions remembered that in the cold, Pre-Boreal centuries after the ice, Doggerland had been tundra where the disappearing megafauna had once roamed. Man-made objects such as worked flints were found in coastal waters, but perhaps most informative was the moorlog itself.

In the 1930s, the science of palynology, the study of ancient plant pollen preserved in geological or archaeological deposits, was in its infancy. But much intrigued by the discovery of the spear point on the Leman and Ower Banks, a group of brilliant young researchers from Cambridge University persuaded Captain Lockwood to take them aboard the SS *Colinda* and return to the findspot. There, in the summer of 1932, two botanists, Margaret and Harry Godwin, retrieved samples of peat or moorlog and took them back to their laboratory for analysis.

Their findings were sensational. Hidden in the midst of the peat, the point had been used in freshwater and not saltwater, and the samples of ancient pollen matched those taken on dry land, on the shores of the North Sea. What the Godwins confirmed was not only the existence of a submerged subcontinent, what had for many millennia made Britain a European peninsula, but some sense of what its landscape had looked like. Doggerland had been a vast expanse of low, rolling hills and river valleys where hunter-gatherers had been able to enjoy a wild harvest of nuts, fruits, roots and animals. Between the Pennines and Poland, a wide northern European plain had stretched out, a habitat where very similar prehistoric cultures flourished.

The weather was kind to Doggerlanders. In what is known as the Boreal period, between *c* 8800 BC and *c* 5800 BC, it was generally dry and temperatures were slightly higher than modern norms. By the end of this benign phase, the long European plain had been carpeted by the wildwood, a green canopy of oak, elm, elder and alder that stretched towards every horizon, a temperate jungle where the sun only penetrated when it was thinned by the falling of the leaves in autumn. Most tracks were made by animals and rivers, lakes, mountains and wetlands opened the only breaks in the wildwood. Hazel trees were common in Britain and there is evidence of a degree of woodland management by our ancestors. Hazelnuts were nourishing and they could be roasted (thereby improving their flavour) and preserved for the hungry months of the winter.

While the Godwins' findings were sensational, proving the existence of a forgotten subcontinent, Doggerland did not escape from academe to become part of our history, at least in the minds of ordinary people. Despite the fact that in the 1970s a Dutch archaeologist, Louwe Kooijmans, reported finds of bone tools dredged up from the bed of the North Sea off Holland, at the Brown Banks, south of the Leman and Ower, there was little or no sense of popular interest, or even knowledge of Doggerland. Perhaps this was because its landscape remained only sketchily understood, submerged under the chill, grey waves of the sea.

It was not until 1998 that the first speculative reconstructions and maps began to appear. Professor Bryony Coles of Bristol University produced a sequence showing the greatest extent of Doggerland immediately after the last ice age, when it was possible to walk from Shetland to the Norwegian coast, and its final centuries as its remaining islands were inundated. But it was not until data from an unlikely source was analysed that these speculations were made real.

Doggerland and Neanderthal

In 2009 a fragment of a fossilised Neanderthal skull was found by Dutch archaeologists – but not in Holland. It was dredged up from the bottom of the North Sea, off the coast of Zeeland. Many fossilised bones have been brought to the surface, but this was a first. Stone tools made and used by Neanderthals and the bones of prey animals like mammoths had been found and their discovery strongly suggested that these people had hunted in Doggerland but no traces of their skeletons had been found. However, when a sharp-eyed amateur palaeontologist noticed the piece of fossilised bone in the waste of a shell-fishing dredger, suspicions were confirmed. The Max Planck Institute in Leipzig analysed the fragment and found that it came from the skull of a young man, and like most Neanderthals, he appeared to have lived on a diet that consisted mainly of meat. He hunted the plains and forests of the great subcontinent some time around 38,000 BC. The young Neanderthal may have had company, perhaps even rivals. Our species, *Homo sapiens*, had arrived in Britain by that time, crossing Doggerland to reach the peninsula.

In 1964 oil companies began to explore the bed of the North Sea in an effort to find new deposits of oil and gas. They were quickly successful and in 1966 a well was sunk on the Leman Bank, not far from where Pilgrim Lockwood had found the antler spear point. Gas was discovered and a new and nearby source of fossil fuels began to be exploited.

In search of more deposits, the oil companies conducted extensive seismic surveys of the sea bed. To collect readings, survey ships trailed cables behind them, crossing and recrossing areas in a grid pattern. The cables transmitted pulses of sound waves through the water, and these were reflected by contrasting geological features on the sea-bed. Once the readings were recorded, vividly coloured 3D models of not only the bed but what lay immediately below it were made. The oil companies' geologists then analysed these to find places where exploratory drilling might be most likely to succeed.

But they also discovered something else, something they did not expect.

Led by Professor Vince Gaffney of Birmingham University, researchers realised that the oil companies were also inadvertently compiling maps of Doggerland. Ancient river courses, deltas, lakes and hills were all showing up in the 3D data. A famous Birmingham professor was memorialised when his name was given to a new river found in Doggerland. Hydrogeologist Fred Shotton had been parachuted into occupied France to map the invasion beaches for D-Day, and his bravery was remembered when the course of the Shotton River was traced across the bed of the North Sea by his successors at Birmingham.

Modern rivers also ran through Doggerland and the new data showed a watershed ridge running from East Anglia in an arc to the Hook of Holland, both areas of dry land being at a much higher altitude in prehistory than they are now. To the south of the ridge, the Thames joined with the Rhine, the Meuse and the Scheldt before turning south-west to flow wide and slow into what was to become the English Channel, while to the north, the Weser and the Elbe emptied into the North Sea not far from the Norwegian coast. Near the middle of Doggerland lay an inland freshwater sea, now the deeps known as the Outer Silver Pit.

In 2012 a remarkable exhibition was mounted at the Royal Society in London, by a team drawn from the universities of St Andrews, Dundee, Aberdeen, Birmingham and Wales Trinity St David. The story of Doggerland was told in more detail than ever before as a wealth of new data was revealed. Having analysed the flora and fauna of the drowned landscape, the team of scientists had been able to calculate what they called its carrying capacity, the number of people Doggerland could have supported. Tens of thousands was the answer, and more, the researchers believed that the lost subcontinent was no less than the heartland of prehistoric Europe.

Naturally, the ceaseless action of the sea over many millennia has eroded much of the archaeology, but nevertheless very intriguing finds have come to light. A mass grave of mammoth bones suggests several interpretations of human activity in the early period of Doggerland, when the subcontinent was at its greatest extent and the northern parts were tundra. Perhaps there was a mass kill, perhaps the bones had been gathered there for some ritual purpose, or maybe to build a structure of the sort raised in Siberia much later.

Possible human burial sites have been found under the sea, and these may date to a later period in Doggerland. Undersea archaeologists have come across arrangements of standing stones that are difficult to explain through natural causes. But perhaps the most exciting hint of a religious and cultural life is a series of mounds surrounded by ditches. In a conjecture that far outruns such slight evidence, it may be that Doggerland was the place where the habit of building henges originated. At first these were not the sort of monumental stone structure typified by the most famous of all, Stonehenge, but rather a simple digging of a roughly circular ditch and making a bank from the upcast. After 3300 BC henges begin to appear in eastern Scotland, earlier than anywhere else in Britain. Is there a link with the submerged mounds and ditches of Doggerland? The subcontinent had almost certainly drowned more than a millennium before, but perhaps a tradition of belief had been passed down through the descendants of refugees who had found the sanctuary of dry land in Scotland.

However all that may be, significant early aspects of Britain's prehistory were contemporaneous with Doggerland and undoubtedly linked with its culture. At Howick on the Northumberland coast near Alnwick and at East Barns in East Lothian, two remarkable structures have been identified. Reconstructed through a careful reading of the depth and angles of their postholes, these were very substantial houses made from the trunks of trees. Rammed and chocked into wide postholes, they were canted inwards to form a tipi shape and lashed together at the apex. When the gaps had been filled by turf or other vegetation and all made as weathertight as possible, these houses would have made snug accommodation for about eight people. But their wider significance is that far from being the sort of insubstantial shelters thrown up by nomadic hunter-gatherers, the Howick and East Barns houses were heavy investments of effort, time and technological skill. And in turn, the houses implied a sense of ownership or at least customary rights over the surrounding land

65

and sea-shore. All of that hard work speaks of permanence and settlement.

In the summer of 2012 a third house built in a similar style was found near the shores of the Firth of Forth, 30 miles west of East Barns. As work on the approach roads to the new Forth Road Bridge intensified and the huge earth-moving machines reached a field near the village of Echline, near South Queensferry, excavators came across another set of tell-tale postholes around an oval pit. It measured seven metres in length, was half a metre deep and surrounded by holes that, like Howick and East Barns, implied a conical tipi shape. Many flint arrowheads and other blades were found and the remains of several internal hearths that would have kept the house warm if a little smoky as the chill winds of winter blew off the firth.

In addition to around 1,000 flint artefacts, large quantities of charred hazelnuts were turned up, and since these are organic and grow in only one year, they could be more precisely carbon-dated. It seems that the house at Echline was built around 8240 BC, at about the same time as those at East Barns and Howick. It was larger and archaeologists believe that it was winter quarters for a band of hunter-gatherers, probably a family group. All of the three houses may have been occupied in wintertime since they were built near the ancient coastline, an excellent and reliable source of year-round food. Beaches and their rock-pools could be combed for shellfish and crabs while the sea itself could be fished. It may be that seal hunts were a regular winter activity.

Hunter-gatherers needed wide ranges, especially in winter when the ability to collect large quantities of firewood was vital and it may be that the people who lived at Echline were in contact with their neighbours at East Barns. Perhaps the edges of their ranges met somewhere east of Edinburgh. In any event, all of them must have been aware of Doggerland, and before 8000 BC it still stretched over much of the northern North Sea basin. There appears to have been a wide and slowly growing gulf between the Scottish and Northumberland coastlines (even though the latter reached further east than they do now) and the western shores of north Doggerland. It may have been possible on a clear day to make out the faint horizon of the Dogger Hills across the sea.

What cannot be doubted are the cultural links between prehistoric peoples who lived so close to each other, and houses very similar to those found at Echline, East Barns and Howick will have been raised on coastal sites around the subcontinent. It is important not to make a distinction between British prehistory and the prehistory of Doggerland. They were the same people and DNA shows that we are probably their descendants. Contact depended on mobility before the modern age, and given the coastal locations of the three big timber houses (and more that will never be found) and the fact that Doggerland was patterned with many river systems and lakes, boats will have been common and their technology and the skills needed to row and paddle them well understood.

A more vivid sense of life around 8000 BC is hinted at in a remarkable place in Yorkshire. It seems that not only did hunters engrave and paint the animals they depended on, they also danced in imitation of them. At one of the earliest sites of human settlement yet found in Britain, 20 sets of deer antlers have been found, and they were not so much trophies as objects of great and perhaps magical importance. Dating to about 8700 BC, only a few centuries after the end of the Cold Snap, the antlers were found at Starr Carr, not far from Scarborough.

The first archaeological dig ever to be carbon-dated, it lay on the shore of an ancient lake, what prehistorians have called Lake Flixton. Clearly possessed of sufficient carpentry skills to split logs and build a wooden causeway through the reed beds by the lake, the people of Starr Carr also made an area of hard standing on the muddy flats by the shore from birch boughs and brushwood. It seems that this place was a hive of prehistoric industry.

Debris from an estimated 14,000 flint tools has been found but far more surprisingly, 220 artefacts made from the antler bone of red deer have been discovered. This is a huge quantity, far more than the proceeds of hunting by one band living by the lake. It looks very much as though the people of Starr Carr were specialists, craftsmen or women whose skills in carving useful objects from antler bone were well known and prized. Production of this sort on this scale clearly implies trade, or at least exchange, and a prehistoric economic network that was more than local.

The most mysterious objects were the 20 sets of deer antlers. Known as frontlets, these were the foreheads of mature red deer with the antlers left attached, but a craftsman had bored holes through the bone of the skull so that things could be threaded and tied, probably under the chin. And the bone of the forehead had been shaped in order for it to fit more comfortably on a human head.

There can be little doubt that these were headdresses, but their use and purposes are less certain. The likelihood has to be that these were ritual. Several hunting cultures recorded in the modern period, such as the Plain Indians of North America who were so dependent on the great herds of buffalo, danced in an act of reverence while wearing headdresses made from the skulls and horns of bulls. It may well be that the people of Starr Carr, who probably traded in antler artefacts as well as hunting red deer, paid a similar homage as men tied the things under their chins.

If the people of Starr Carr did indeed have spiritual beliefs then these are likely to have developed over time, and three millennia after the deer men danced, whatever the nature of their gods, those who lived on the Yorkshire coast or along the north-eastern coasts of Britain or in Doggerland, may have thought that their own rituals had ordained the end of the world.

Deep in the darkness of the Storegga Trench, an undersea chasm lying in the North Atlantic between Norway and Iceland, a seismic disaster was waiting to happen. The Storegga is what geologists call a subduction trench, and some time around 5100 BC it suddenly stirred, opened and caused 3400 cubic kilometres of sand, gravel and rock to fall into it. Along a 290-kilometre length of coastal shelf, an area the size of Scotland broke off and collapsed into the void. This undersea earthquake caused bafflement and then catastrophe. The coastal peoples of Norway, eastern Britain and Doggerland will have watched awestruck as the sea first retreated very quickly, far further than the low-tide mark. And then they will have heard the distant rumble of a giant wave, a tsunami, racing towards them. Roaring across the surface of the sea at 600 miles an hour, an 8-metre high wave smashed into coastlines, killing many, destroying settlements, inundating a vast area.

The Great Tsunami

After the convulsion caused by the underwater seismic activity known as the Storegga Slide, researchers now believe that the tsunami it caused raced across the North Atlantic in only three hours. This meant a 3-metre-high first wave (others followed) travelled at an unimaginable 600 miles an hour before it smashed into coastlines. On the nearest, the Norwegian coast, the tsunami had a maximum run-up of 10 to 12 metres and on the eastern coasts of Scotland run-up heights exceed 3 to 5 metres. The total length of inundated coastline was more than 600 kilometres. Sand deposited by the tsunami allows scientists to make calculations, but the impact of the wave-front on the submerged coasts of Doggerland can only be conjectured. However, one telling detail has emerged from the analysis of the sand deposited by the tsunami. Twigs and fish-bones found on the Norwegian coasts suggest that the Storegga Slide took place in late autumn, a time when stores were being laid in for the winter. The tsunami probably destroyed much of this supply of food, causing starvation amongst those who survived the initial devastation caused by the wave-front. Doggerland may have been completely abandoned after this catastrophe.

For millennia Doggerland had been slowly sinking and all that may have remained of its rolling landscape and river valleys above the waves was Dogger Island, the northern range of hills that in an instant had become the Dogger Bank. South of it there may have been a scatter of smaller islands not directly in the devastating path of the tsunami, perhaps on the Leman and Ower Banks or the Brown Banks. By *c* 5100 BC the sea had broken through to form the English Channel and Britain had at last become isolated.

Since the sea had begun to rise after the end of the last ice age, sometimes at a rate of 1.5 metres in a century, bands of hunter-gatherers had been seeking dry land. This process will have been accelerated as the lands to the north bounced back after the weight of the ice was lifted. Refugees will have set out in their boats both east and west and DNA supports the sense of a prehistoric migration to British and German coasts. There is a common Y-chromosome marker, which only men carry, labelled as S24, and it is common in Lower Saxony, the valleys of the Weser and the Elbe. It is conceivable that S24 originated in Doggerland – it has been described recently as the heartland of prehistoric Europe and markers certainly originated in much less populated and more unlikely places – and as the coastline shrank back in the east, Doggerlanders moved up the Elbe and the Weser. Almost 10% of modern Lower Saxons carry it. And the marker also travelled west. S24 is common in England, where 4.8% of men carry it, around 1.3 million. It is much less frequently found in Scotland and Ireland.

After the North Sea and the Channel separated Britain from Europe, S24 certainly kept making crossings, most notably in the period of Anglo-Saxon settlement from the fifth to the seventh centuries and even later. But the marker is old, and some of those who carried it certainly walked across the plains of Doggerland. It represents a striking and ancient link, a survival from a drowned subcontinent.

Another early marker harks back to the deer men of Starr Carr, and it too was probably present amongst the hunters and gatherers of Doggerland. In England S185 is significant at a frequency of 0.6%, or 162,000 men. It is also old, dating to at least 6000 BC, and it approached Britain from the east, almost certainly through the valleys of Doggerland. In Ireland, the frequency reaches 2.4% of all men, and this may well be because of comparatively less dilution by new markers arriving in later history. One thing is certain, and that is the direction of travel. The men who carried these DNA markers entered Britain from the east, some of them on foot.

By 4000 BC the wildwood carpeted most of the new island and its tiny population of hunter-gatherers had adapted to its seasons, its flora and fauna. There were fewer animals to hunt since the disappearance of the wide grasslands and the great herds of migrating herbivores. But if the hunters found less prey browsing the leaves and herbage of the forests, there were more hazel and fruit trees, more bushes with berries, more fungi, more birds, more eggs and more roots. Ranges may have been more restricted and movement through the tangle of the greenwood slower, but there is a sense of a life and a habitat that changed only gradually. Although they could not know it, those who lived in Britain at the end of the fifth millennium BC were the last of their kind. They stood on the brink of a revolution, arguably the greatest in human history. Abruptly and emphatically, farming would change all their lives forever.

4

Day In, Day Out

✹

O N T H E O T H E R S I D E of a causeway that divides two lochs on the Mainland of Orkney lies a site that has turned Britain's prehistory upside down, and set an enormously complex problem for geneticists. Between the Harray Loch and the Stenness Loch stand two monumental stone circles, the Ring of Brodgar and the Stones of Stenness. Close by is Maes Howe, a stunning stone-built tomb chamber and in the 1980s a prehistoric village was found at nearby Barnhouse. This impressive landscape of ancient remains was thought to be the focus of religious and ceremonial life on Orkney in the third millennium BC. But startling discoveries that began to come to light after 2002 have forced archaeologists and historians into radical revision. The great stone circles, the tomb chamber and the villages, including nearby Skara Brae, majestic though they undoubtedly are, turned out not to be central but peripheral, monumental adjuncts to a site that had been lost and forgotten for millennia, a place that would reveal wonders and that would demand that Britain's early history be rewritten.

When visitors come in their thousands to marvel at the huge Stones of Stenness each summer, they generally get back on the tour bus and are taken across a short causeway to park beside the Ring of Brodgar, only half a mile to the north-west. For many years, they passed unknowingly through a mystery, an intensely sacred place, a precinct where building began before the stones were raised at Stenness and Brodgar, the site of the most remarkable remains of prehistoric Britain and the centre of the ritual landscape between the two lochs.

Behind Lochview, a white house that stands on the left-hand side of the road as the buses approach the Ring of Brodgar, there rises what was thought to be a natural grassy hillock, on what is known as the Ness of Brodgar. When the stone circles and the other monuments were awarded World Heritage status by UNESCO in 1999, archaeologists were moved to consider the area around the famous sites. The non-invasive technique of geophysics, ground-penetrating radar, was used on the hillock behind Lochview – and they showed that it was not a grass-covered glacial moraine but entirely manmade. The sensitive radar picked up the lines of buried walls, of pathways and the outlines of large buildings. It was a stunning discovery and work to excavate began almost immediately.

As archaeologists dug a pattern of test pits on the Ness, extraordinary evidence began to be revealed. Two massive walls had been built across the isthmus from the shores of one loch to the other. Each wall was found to be more than 100 metres long and 4 metres high, and made from beautifully cut and fitted drystane flagstones. Their positioning not only had the effect of controlling access through the isthmus but they also protected and screened a precinct, an inner sanctum. For behind the Ness of Brodgar's huge walls lay something unique in prehistoric Europe, a large temple complex of unparallelled sophistication, colour, beauty and mystery.

Since 2002, archaeologists have uncovered the massive foundations of 10 temples. The largest is 25 metres in length and 20 metres wide. Probably made from split stone slates, its roof may have sheltered a circuit of walkways under the eaves. From the outside, this huge temple, the largest non-funerary structure yet found in Europe, must have been awe-inspiring. But inside, the space was small, suggesting that only an elite were permitted to pass through its portals.

In 2010 the excavators came across something that changed the perception of all prehistoric structures. The temples of the Ness of Brodgar were painted. A stone slab was found with traces of red, yellow and orange pigment still on it. Probably extracted from iron ore and mixed with a fixative such as animal fat or egg whites, paint was applied not to walls but to individual stones. The effect must have been like a checked cloth or even some tartans. These discoveries supplied hard evidence for what had been long suspected, and they transformed the way in which stone circles and other weather-washed prehistoric monuments are seen. Instead of a sombre grey, these structures were probably vividly decorated with patterns of colour. Just as the Greeks and Romans painted their temples and statues, so did the builders of the Ness of Brodgar, but they mixed their pigments long before the Parthenon or the Pantheon were raised, before even the Pharoahs of Egypt commanded the building of the pyramids.

Archaeologists believe that the foundations of the massive walls were first laid down as early as 3300 BC and perhaps even before then. Two or three centuries later, the first oval buildings in the temple complex were begun and the Stones of Stenness erected on the flatter ground to the south-east. The large building known as the Cathedral was built around 2600 BC and at the same time the Ring of Brodgar rose to the north-west. All of this work was undertaken well before the great stone circle was laid out at Avebury in the south of England and long before the massive bluestones were dragged to Stonehenge some time around 2000 BC.

Henges appear to have been invented in Orkney. The term is somewhat deceptive, deriving as it does from Stonehenge and bringing immediately to mind the great stone lintels of that singular structure. In fact, henges began as a simple enclosure made by digging a broadly circular or oval ditch and piling the upcast on the outside to make a screen. The deeper the ditch, the higher the outer bank. Why these enclosures were dug and what took place inside them will probably remain forever mysterious. Any detailed deductions derived from archaeological evidence are bound to be highly speculative, like trying to reconstruct what happened at mass from looking at the ruins of a medieval abbey. But some basic observations are possible, in fact they are necessary if any sense of the life of this extraordinary, innovative society on prehistoric Orkney is to be brought even a little way out of the shadows.

The banks and ditches of henges were intended to create a place apart, somewhere more sacred, more important in the landscape, an enclosure where ceremonies of some unknowable sort probably took place. And this place apart was screened by a bank, implying that it was not open or accessible to everyone. This emphasis on exclusivity and secret mysteries in religious practice lasted long into the Christian era when until recently churches were divided by rood screens, and beyond them, unseen by the laity, priests enacted the rituals of the mass in a language very few understood.

The massive walls at the Ness of Brodgar appear to have demarcated sanctity, created an inner sanctum. This sense of outside/inside runs through all of these early monuments. And the notion that inside the holy of holies only an elite group could practise whatever rites were at the core of prehistoric religious belief is underscored by the architecture of the Cathedral at the Ness. Imposing from the outside but very small and restricted inside.

As the complex of temples took shape, work also went on immediately to the south, where the magnificent circle of the Stones of Stenness was laid out. The ditch was cut through rock, a tremendous labour with only stone and bone tools to hand, to a depth of two metres and a width of seven. The upcast was piled on the outer edge. The sense of a boundary was further emphasised by the fact that the ditch probably lay below the water table and, once complete, it will have filled up like a moat.

Inside the circle, sockets were dug to hold 12 standing stones, although it appears that some time around 3000 BC only 11 were erected. Six remained in the early nineteenth century but even these survivors were almost lost. Captain W. Mackay, a neighbouring farmer and a recent incomer to Orkney, grew irritated at people crossing his farmland to visit the stones and in December 1814, he and his men smashed what was known as the Odin Stone and destroyed another. Local people were furious and Mackay was physically restrained from causing further damage. The four stones still standing are extraordinarily impressive, the tallest being more than five metres high. All of them are less than half a metre thick and they seem to soar like prehistoric spires.

The sole entrance to the circle at Stenness faces the site of the ancient village at Barnhouse while around it stand other sentinel stones. The Watch Stone is the largest and it guards the causeway leading to the Ness of Brodgar and the Ring of Brodgar. It is thought that in the fourth and third millennia BC the freshwater of Harray Loch ran into the saltier water of Stenness Loch at that place and the rivulet was crossed by a series of stepping stones.

On the isthmus beyond the crossing, the Ring of Brodgar rose some time around 2500 BC. At 104 metres, its diameter is more than double that of Stenness and it was originally encircled by 60 stones, of which 27 remain standing. Brodgar is not technically a henge since there appears to have been no bank and its huge scale seems to signal either a significant increase in population or a change in belief, a more inclusive celebration of rituals. Perhaps the exclusivity of the Cathedral at the Ness and Stenness was waning. In any case, Brodgar was still the product of intense and co-ordinated labour undertaken by many people. The stones stood inside a circular ditch which again was hacked out of the living rock, this time to a depth of 3 metres, but there seems to have been no screening bank. Perhaps all those involved in ceremonies could gather inside the circle. There was certainly room to accommodate many.

The Ness and both of the great stone circles were not raised around the isthmus between Lochs Harray and Stennes by chance. Such immense effort on difficult ground where the rock was only just below the surface, and the hundreds of thousands of man-hours involved demanded that the location was very carefully considered. Certainly it is beautiful. When the builders of the great walls on the Ness looked up from their work, they gazed upon a dramatically beautiful landscape – but it was not the drama of mountains or the crashing breakers of the mighty Atlantic. Instead it was a quieter vista of fertility, cultivation and stock-rearing that they saw. The two lochs lie in the middle of the Mainland, by far the largest of the 70 islands in the Orkney archipelago, and much of it is still very fertile farmland. In prehistory men and women looked on that landscape, knowing that it fed and nurtured them and their families. The vistas around the lochs are bounded on every side by a ring of low hills, and there is a sense of a green and productive amphitheatre of plenty. Perhaps the positioning of the temple complex and the stone circles in the centre of that bountiful landscape was a means of giving thanks or of appeasing the higher powers that enabled all that plenty. The circles are open to the skies (and very few could actually enter the Cathedral), they appear to have been aligned on astronomical principles and the direction of worship and focus was probably skyward.

While the precise nature of the powerful beliefs that caused the monuments to be built can never be known, the placing of the stone circles either side of the Ness of Brodgar does whisper some hints of the pattern of events. Sophisticated, unique, impressive, and first to be constructed, the temple complex was the focus. By 2500 BC, when both the Cathedral and the Ring of Brodgar had been completed, the stone circles may have served as gathering places where the people of the West Mainland and the East Mainland approached the sacred centre of the island. Processions to the massive walls may have followed. It seems inconceivable that people who gathered for some sort of religious purpose moved silently through the landscape. Archaeologists have not only found prehistoric musical instruments in the delicate shape of bird-bone flutes, rattles and drums (here and in several sites across Europe), there must have been singing or chanting.

As well as the life-giving land, death appears to have been at the heart of ceremony. Orkney is dotted with many chambered tombs, 76 in all. Built to resemble houses of the dead, some were found to contain the disarticulated bones of many people. The first burials were placed in the Tomb of the Eagles on South Ronaldsay around 3150 BC and it was closed more than seven centuries later, about *c* 2400 BC. On the island of Rousay there were 13 chambered tombs and archaeologists noted a remarkable continuity in that there are exactly 13 modern farms (or tunships). In general the distribution of these houses of the dead is not clustered in a specially sanctified area but each appears to be attached to a parcel of good farmland. It seems very likely that the tombs also served as territory markers.

While the precise rituals of burial are now lost in the darkness of the prehistoric past, it seems that the act of interring ancestors in the ground worked by individual kindreds was itself an affirmation of ownership. By planting the bones of those who had toiled to keep the soil fertile and its animals healthy and productive in a specially constructed house of the ancestors, it was as though their tenure continued unbroken by death. The presence of their bones made it the kindred ground. All sorts of rights were asserted thereby. In addition, the piles of bones and skulls in the chambered tomb had a sense of returning to the land those whose lives were made possible by its fertility – ashes to ashes, dust to dust, as modern rituals have it.

When work began on the huge central complex around the Ness of Brodgar, it is as though power also shifted and centralised. Local tombs were still used and maintained but the scale of what was built between the lochs speaks of great authority, perhaps of priest-kings or -queens of Orkney, those who could command and co-ordinate these logistically sophisticated operations that involved many people and materials. Chambered tombs could be built by communities but the temples and stone circles needed all of Orkney's resources.

Who were these people, these innovators and visionaries, these priest-kings or -queens? The beginnings of an answer lie at the Knap of Howar. In the 1920s the coast of Papa Westray eroded sufficiently to reveal the remains of two very early farmhouses. The site was radiocarbon-dated to *c* 3600 BC. Grains of wheat and barley were found as well as evidence that cattle and sheep were reared and routinely slaughtered for their meat while still young. Apparently the bones of pigs suggested a larger than normal breed, animals much closer to the wild boar. It was an emblematic find. The name of Orkney derives from these aggressive creatures. It was first recorded *c* 320 BC by Pytheas when he wrote of his visit to Orkas, the islands of the Boar Kindred. It may have been a totem animal revered for its various qualities, for others were noted near at hand. On early Roman maps, in part derived from Pytheas's records, Caithness was the land of the Lynx Kindred, Sutherland the home of the Raven People, and to the west lay the lands of the Caereni, the Sheep Folk.

What the finds at Knap of Howar signified for British and Orcadian prehistory was something simple, a very early example of houses built by farmers, by people who may have continued to supplement their diet by hunting and gathering, but whose focus was on growing crops and rearing domesticated animals. The site of the ruined stone buildings on the sand-blown shores of Papa Westray is one of several monuments to the greatest revolution in our history, the coming of farming. Despite its immense importance, the arrival of this new way of life in Britain is poorly understood and a matter for much speculation. But DNA analysis can at last shed light on who these people were and where they came from.

It was a revolution in a very meaningful sense and it began some time around 8500 BC in the Near East, the arc known as the Fertile Crescent, from Iraq through Syria to the Levant. There hunter-gatherers had managed their ranges, encouraging the growth of fruit trees and berry bushes, trying to ensure a continuity of supply. But at some point in the ninth millennium BC, stands of fruit trees became orchards, gardens were planted instead of being the semi-accidental product of self-propagation, and crucially, wild grasses were cultivated as cereals. Alongside these developments, manageable species of animals were domesticated for their milk, wool, hide and meat.

Lethal Zebras

When European horse breeders settled in South Africa in the 1600s, they attempted to domesticate zebras. It proved impossible. Despite their attractive stripes and noble heads, they are extremely bad tempered and will bite anyone attempting to handle them. And not let go until the handler is dead. In zoos, more keepers are injured by zebras than tigers. They cannot be lassooed or have a halter placed on them because, being prey animals, they have better peripheral vision than horses and simply flick their heads away from whatever comes at them. Zebras are an excellent example for illustrating why only 14 species of 148 larger mammals have been domesticated. The favoured 14 passed 7 tests. First they needed to be large enough to produce milk, meat, wool and perform work, usually traction, but not so large that they could not be led or housed. They had to eat a diet that could be supplied by human beings, such as hay. They needed to grow quickly and produce many young. They could not have a nasty disposition. They had to be able to breed in captivity, organise themselves in hierarchies and not panic when confined to enclosures. Sheep, goats, small cattle, pigs and 10 other domesticated species passed these tests, while 134 did not.

Arguably, it was the production of primitive wheat and barley that had the greatest impact because it transformed child-rearing. The populations of hunter-gatherer societies rose only very slowly for two compelling reasons. As these bands moved around their wide ranges, relocating from summer to winter camps, going on seasonal hunting expeditions, they needed to be as mobile as possible. That meant only one baby or toddler could be carried along with the other gear needed. And in an age before contraception, another factor came into play. As has been noted, infant teeth could be too soft to deal with the hunter-gatherer diet and in order to take in enough protein to grow, babies and toddlers almost certainly breast-fed for much longer, perhaps only being weaned as late as four or five years old. During this lengthy period, breast-feeding mothers were usually infertile.

The short fertile lives of most women were another factor that inhibited the growth of populations. There is evidence that women in prehistory began their menstrual cycle later, maybe at the age of 13 to 15, and many surveys of surviving skeletons report that most people died relatively young, with few of them reaching their 30s. Over such a short time, most women will have given birth to only three or at most four babies, not all of whom will have reached adulthood. The production of cereals changed this cycle radically.

When the ears of wheat were dried, and sometimes charred, they could be mashed into a nourishing porridge with animal milk or water. Not needing to be chewed, this could be fed to infants and they thrived. This in turn led to earlier weaning – and an explosion in the prehistoric population. A Scotsman might remark that porridge changed the world – and he might be right.

As the population grew after *c* 8500 BC in the Near East, pressure on the land and on communities built up. Farming led to a powerful sense of the ownership of land as those who had expended great labour in creating gardens and small fields insisted on their rights. That in turn forced a calculation. In the new world of farming, what was the carrying capacity of the land, how many mouths could its produce feed? When the birth interval halved from four to five years to two or three, that led to a rapid increase in numbers. As more and more land was brought into cultivation, those whom it could no longer support were forced to move and the techniques of farming began to ripple westwards from the Fertile Crescent towards Europe.

Escargot Cargo

For generations scientists have struggled to understand why Ireland shares some plant and animal species with the Iberian Peninsula but not with Britain or the rest of Europe. Land snails blazed a DNA trail. Researchers knew that the *Cepaea nemoralis*, snails found on the western coasts of Ireland, share an unusual trait with those found in the southern foothills of the Pyrenees. Both have a distinctive white lip on their shells. But then scientists from Nottingham University took DNA samples from species of snail all over Europe and found that the Irish snails and their Iberian cousins shared a variation in one gene that makes them different from all other European snails, giving them the white stripe. The most plausible explanation is that migrants sailed north with *Cepaea nemoralis* around 8,000 years ago with the intention of cultivating them as a food source. If this conjecture is correct, it could be argued that the escargot was the first animal to be farmed in Europe.

Ferriter's Cove is an elemental place, one of the westernmost edges of Europe, a beautiful sandy bay on the shores of the Dingle Peninsula in south-western Ireland. It faces the apparently limitless Atlantic but, some time before *c* 4350 BC, it was landfall for people sailing from the south. Following the chance find of part of a flint knife, an object that spoke of early farming communities, archaeologists from the University of Cork dug into the sandy soil and found something very surprising, something that pulls the flow of our history away from the south-east of Britain, like the great building projects on Orkney. Cattle bones and a sheep's tooth came to light and the former dated very early, to *c* 4350 BC, the remains of the first domesticated animals so far discovered to have come to Britain and Ireland. Three other Irish sites have since produced similar finds and it seems that these first farmers, the people who brought these cattle, and probably sheep, sailed from the south, approaching Britain and Ireland by an Atlantic route, carrying in their boats the seeds of a revolution.

By the time these new people pulled their boats up above the tideline at Ferriter's Cove, Doggerland had been submerged and perhaps only Dogger Island remained, close to the point of being inundated. The North Sea and the English Channel had isolated Britain at last and Ireland had been an island for even longer. The first farmers brought their ideas, their seed corn and their animals in boats, and each crossing must have been hazardous. If these were curraghs, hide boats made from fragile frames of willow and other whippy woods, they will have made awkward transport for animals. Bobbing on the tops of the waves, probably without a sail, propelled only by oars, currents and tides, these craft were not ideal as animal transports. And yet they must have made the trip, and with animals on board. Sheep have no wild progenitors in Britain or Ireland and must have been imported by these new people. Wooden boats only appear in the archaeological record after 2000 BC because their construction, for which the splitting and the making of planking was central, depended on metal tools. Did these first farmers put down wooden decking below the thwarts, a flattish floor where hobbled animals were laid on their sides? The sudden movement of a frightened cow could have been disastrous, capsizing the boat. Perhaps they brought only younger and smaller animals such as weaned lambs and calves. However it was managed, there must have been many successful voyages – for they came, these intrepid people, and they changed the landscape and our culture forever.

But who were they? And where did they come from? DNA data strongly suggests that most of the first farmers were men, and that they sailed to Britain and Ireland from two directions. But their Y-chromosome markers have left only a faint trace, an identity that began to come to light more than 20 years ago.

A chance discovery high in the Italian Alps supplied surprising evidence of a disease that had transferred from animals to human beings. In September 1991 two hikers discovered a human corpse in the Tisenjoch Pass in the Ötztal Alps. His body had been miraculously preserved, mummified in the ice of the mountainside. Quickly dubbed Ötzi, the man had clearly died a violent death since his left arm had been raised, freeze-framed, fending off a blow. An arrowhead was found lodged in the soft tissue of his left shoulder. Perhaps a shepherd who tended his flocks in the high summer pasture, certainly a well-equipped huntsman, Ötzi lived some time around 2300 BC, at least two millennia after farming had arrived in central Europe. And he carried tell-tale signs of the a way of producing food which by that time was two millennia old.

Ötzi had Lyme Disease. Carried by ticks found on rodents, deer and sheep, it can be very debilitating, sometimes fatal, and the nature of its transmission has only become understood in the last 20 years. The Iceman's DNA also revealed that he was lactose intolerant, or unable to digest animal milk, such as that of cows, goats or sheep. Despite the fact that he lived in the new age of farming and may have been a herdsman, this was probably not as much of a handicap as it might at first seem. Accumulating evidence of prehistoric cheese-making has been found by archaeologists across Asia Minor and into Europe. Milk residues have been detected in pottery dating to the seventh millennium BC in north-west Anatolia, and more tellingly, potsherds with small holes deliberately bored into them have turned up in Europe and Britain. These were almost certainly cheese strainers, and their use will have made life and lactose more tolerable for Ötzi.

Cheese is made by separating the more solid and fatty curds from the whey, which contains most of the lactose. Cheddar cheese, for example, has only 5% of the lactose contained in whole milk. Intolerance is not absolute and Ötzi could probably have munched a wedge of cheddar without mishap. And a distinct advantage of treating milk in this way is that cheese lasts a great deal longer, especially if it has had salt added to it. Lactose intolerance is now fairly rare in western Europe and its frequency was no doubt much diminished by natural selection. As will be noted later, those who could tolerate the drinking of whole milk probably bred faster and their DNA spread more widely.

Ötzi's Y-chromosome marker is in the haplogroup G and it makes a clear link with the development of farming in Europe. At a site near Treilles in south-west France, archaeologists found what they believed to be a communal cemetery and they succeeded in extracting DNA from the teeth of 29 individuals. The skeletons were dated to around 3000 BC. It appeared that most of the women were descended from the local hunter-gatherer communities of that part of France but many of the men were not. Either incomers or the descendants of incomers, they also carried markers from the G haplogroup. Their DNA placed their ancestors as far east as Cyprus and Turkey. And in a fascinating twist to this discovery, it was found that most of the men were closely related while the women were not. Amongst the male skeletons, sets of brothers and fathers and sons were identified.

The cemetery at Treilles seemed to be whispering the lines of a long-lost story, the outline of an event in prehistory. It may well be that this group of related men were either descendants of incomers from the east, perhaps people who had arrived in the wake of the wave of carriers of the G markers. In any event, it seems that they stayed on the land they were cultivating and that their women came from native communities outside that area. This was another pattern that would repeat across Europe. But what is fascinating is that these men shared their DNA with Ötzi. In western Europe G is most common in Sardinia and northern Italy, where 14% and 10% of the male populations respectively carry it. The marker is also significant in England, where 2.5% have it, but less so in Ireland at 1.5% and Scotland at 1.1%.

Almost certainly, those who carried the G markers sailed to Britain and Ireland in groups. Several boats carrying livestock and seed corn were probably beached at Ferriter's Cove and on many other landfalls. Farming is a co-operative activity, the work of many hands and, unlike hunting and gathering, little can be achieved on the land by an individual. Similarly, there was no point in bringing one or two animals, the most that a sea-going curragh could safely hold. Domesticated cattle, goats, pigs and sheep needed to breed and multiply, and expeditions to colonise new farmland were very likely enterprises undertaken by migrating groups of men. Which would have had an even more immediate effect on population change and growth.

Average estimates of the sparse but stable hunter-gatherer population of Britain after the end of the last ice age lie somewhere between 5,500 and 10,000, but by *c* 4,000 BC it is thought that it increased tenfold, to 100,000. At approximately the same time, the population of Ireland may have risen from 2,000 to 4,000 up to 40,000. When the wave of new people washed over these islands at the farthest north-west of Europe, the effect of gene-surfing, like the founder effect, was disproportionately dramatic.

When the farmers first waded ashore and pulled up their boats behind them, they were not entering virgin territory, land that was there for the taking. While the hunter-gatherer population was small and widely dispersed, its people did enjoy customary rights over large ranges, familiar hunting ground that they will not have wished to give up. An option may have been to adopt the new techniques, a new way of life that needed a smaller area. Did the natives become farmers? It seems very unlikely. Again, DNA offers answers.

When the incoming groups established themselves and carved out their farms and pastures, populations exploded. As cereal production began and the birth interval halved, the carriers of G saw their Y-chromosome marker proliferate very rapidly. If men who carried the older, hunter-gatherer markers had adopted the new way of life on any scale, their Y-chromosome DNA could also have multiplied. But it seems that they did not. Only 1.5% of all British and Irish Y-chromosome markers, principally M26 from the ice age refuges and S185 from across Doggerland, certainly pre-dated the arrival of the farmers *c* 4000 BC, and a further 3.5% may have done.

What underlines this dramatic turn of our prehistory, this cultural and genetic revolution, is the frequency of mtDNA markers – what women pass on to their children. The picture is different. Some women did sail with the new farmers, markers such as K certainly arose in the Near East and around 8% of British women carry it or have passed it on to their children. And the mt marker J, found in around 10% of the population, has been labelled that of the First Farmers. But it is likely that, as they did at Treilles, the incomers took native women as partners and the markers of H and V, those of the pioneers, survived and thrived for that reason.

The new men were vigorous and certainly out-bred the native hunter-gatherers, but did they out-fight them? Evidence for conflict in prehistoric Britain is slight, but it seems likely that it took place. At sites in the south of England, archaeologists have found clues. Burials have been excavated where the skeletons had suffered arrow wounds. Examples were found at West Kennet Barrow in Wiltshire, which was built *c* 3300 BC, a date that suggests that conflicts recurred over several centuries. Similar evidence has turned up at Wayland's Smithy and Ascott under Wychwood. To the south, the enclosures at Hambledon, Hembury and Carn Brea in Cornwall were all attacked some time between 3700 BC and 3500 BC by bands of archers. Crickley Hill in Gloucestershire was also assaulted. While these sites may have been contested by rival warbands of farming communities rather than incoming farmers and hunter-gatherers, it seems that in the fourth millennium BC warfare was widespread.

Perhaps the most spectacular example of the impact made on native populations by incomers took place in relatively recent history, in the Spanish and Portuguese conquests of Latin America. When a lookout shouted that he had sighted land early in the morning of 12 October 1492, his captain, Christopher Columbus, was exultant, vindicated and expectant. The prospect of great wealth as well as a sense of adventure had prompted his five-week-long voyage into the unknown and those hundreds of thousands who followed in his pioneering wake were ruthless in their pursuit of gain. And in the process of acquiring all that gold, silver, sugar and other valuables, they almost obliterated Native American society. Their extermination of the men of the Caribbean islands and Central and South America was so thorough, so overwhelmingly complete in some places that it may be seen as a parallel for what probably happened in Britain and Ireland when the deceptively peaceful process of farming arrived after *c* 4200 BC. But crucially, what took place after Columbus's ships sighted land in 1492 was recorded.

After coming ashore on one of the islands of the Bahamian archipelago and finding the native inhabitants friendly and curious, the three famous ships of the Spanish expedition sailed further west to Cuba and then on to Hispaniola, the island now divided into Haiti and the Dominican Republic. In the decades that followed, carnage took place, a mixture of genocide and plague now reflected in the genes of modern Cubans.

The men of the Ciboneys and the Tainos, the two dominant native groups on the island, were completely eradicated. No traces whatever of any native Cuban lineage were found in a recent survey of the Y-chromosome. Instead, the fatherlines of modern Cubans are now made up of 80% Eurasian and 20% African, the latter a legacy of the centuries of slave trade that followed the discovery of the Americas, something made necessary by the devastation caused by the new colonists. No men were left to work in the plantations and the mines. Traces of the Ciboneys and the Tainos can only be found in the maternal lineages, and 33% of the mtDNA of Cuba is native. The balance comes from Africa at 45% and Eurasia at 22%. This pattern repeated as colonial history rumbled over the islands of the Caribbean and the mainland of the Americas.

After leaving the shores of Cuba, Christopher Columbus spent time as the Governor of Hispaniola, a role that seemed not to suit the talents of an adventurer and explorer. He was dismissed from the post, sent to prison in chains and put on trial, one of the principal charges being that he and his brothers tortured natives and governed through terror. Again, no trace of the indigenous Y-chromosome DNA of the island can be found in the modern population. But perhaps even more dramatic examples of what took place as hundreds of thousands crossed the Atlantic (it is thought that approximately 240,000 Spaniards came in the sixteenth century and *c* 500,000 in the seventeenth) are to be found in studies done amongst the mainland populations of Latin America. Sampling in Colombia showed that 94% of the Y-chromosome markers are European in origin, 5% African and only 1% native. By extreme contrast, a staggering 90% of the motherlines, the mtDNA, is native. An extreme imbalance that tells an unambiguous story of the scale of what happened and one that offers some support for the tremendous impact made by carriers of new DNA almost 6,000 years earlier, in Britain and Ireland.

A study of the ancestral DNA of Brazil published in 2000 produced more detail. In essence, it demonstrated no appreciable difference between the Y-chromosome lineages of Portugal and its vast colony of Brazil. The Y-DNA of the men of both countries looks identical and there was no detectable trace of Native American fatherlines.

In a secondary parallel, Latin America was also colonised by small groups of men in small boats. The 'Catalogo de Pasajeros a Indias' lists those who crossed the Atlantic from Spain to the Americas between 1509 and 1559. By no means exhaustive, it records that only 10% of the 15,000 colonists were women. The balance of 90% of the Y-DNA lineages in Colombia being European broadly agrees with the evidence of the passenger lists. But the preponderance of male conquistadores and colonists was by no means the only demographic factor in play.

Bandeirantes

Meaning the 'followers of the banner', the name described bands of Portuguese slave-hunters who penetrated the interior of Brazil and in essence established the limits of this vast colony even though they were unofficial and unconnected to the state. Between 1580 and 1670, the Bandeirantes often used the tactic of disguising themselves as Jesuits, holding a mass to lure natives out of their settlements. They also used surprise tactics or burned people out of their homes. By the end of the seventeenth century they began to look for gold, silver and diamond mines and the bands were made up of Amerindians, 'caboclos' or mixed-race men and white men. Some became immensely wealthy and the city of São Paulo was a notorious Bandeirantes base.

Disease did the work of conquest – and far more effectively than force of arms. It is estimated that 95% of the pre-Colombian Native American population died in epidemics of smallpox, influenza, measles, typhus and other diseases after 1492. These were brought across the ocean by Europeans and Africans, and the natives they encountered had no immunity. When Hernán Cortés attacked the Aztec Empire 1519, he was at first repulsed. But then smallpox did what his soldiers could not. An epidemic devastated the native population, killing Cuitláhuac, the shortlived successor of Emperor Montezuma. More than 50% of the Aztec population is thought to have perished. After Francisco Pizarro's expedition landed on the Peruvian coast in 1527, smallpox raged amongst the Incas and greatly aided his campaign. This baleful pattern repeated and persisted over the centuries after the arrival of Europeans as colonists all over the world. Following first contact, the peoples of island populations were particularly vulnerable; by the mid nineteenth century Hawaii had lost 90% of its native population to western diseases.

Many of the most devastating killers originated in farming populations. Close and regular physical contact with cows and the ingestion of their nourishing milk encouraged the likes of measles, tuberculosis and smallpox to jump the species barrier and infect human beings. Influenza and its virulent strains have attacked communities all over the world and it appears that in prehistory, the first farmers caught it from pigs and ducks. The carriers of G to Britain and Ireland after *c* 4200 BC may have brought something much more deadly than their genes.

Archaeology and DNA appear to agree on the origins of the vast majority of the incomers to Britain and Ireland after *c* 4200 BC, those enriched for the G lineages. There were at least two major routes taken to Britain and Ireland. The shortest and least hazardous crossing was to sail the English Channel, and many voyages began on the coasts of what is now France, Belgium and Holland. Those who landed at Ferriter's Cove and on the Atlantic shores of Britain set a western course from Brittany, Atlantic France and probably Iberia. The origins of the DNA of both these groups confirm the two approaches. And finds of two different types of pottery also make a secure link. What is known as Linearbandkeramik, a decorated ware, came from the east, while impressed or cardial ware arrived in Britain from the Atlantic seaways.

In this era of migrations Orkney may not have been fought over. Before the early farmers arrived in the period before Knap of Howar was built in *c* 3600 BC, the archipelago was probably not inhabited. Only one tiny scrap of evidence for occupation before that date has ever been found: a charred hazelnut shell at Tankerness was carbon-dated to *c* 6000 BC. And that may have been the deposit of a summer expedition from the mainland of Scotland.

If Orkney was an empty, green and fertile Arcadia, that may have encouraged the precocious culture that built the temples at the Ness of Brodgar and the monuments around them. Other innovations flowed south. Too heavy and fragile for the mobile lifestyle of the hunter-gatherers, pottery, the first man-made product in Britain, was needed by sedentary farming communities for storage. A more sophisticated type known as grooved ware originated in Orkney before being adopted all over Britain.

But it is very difficult to identify the DNA of this remarkable society. Modern Orkney is not populated by a high percentage of men who carry G lineages – in fact they are rarer than in mainland

Britain. Instead, another marker from a much later period seems to have largely supplanted the DNA of the farmers. Scandinavians, the Vikings, began raiding at the end of the eighth century AD and when they settled Orkney, they seem to have carried out a similarly brutal removal of the natives as happened across the rest of Britain *c* 4000 BC.

The four centuries between 4200 BC and 3800 BC saw radical change. The wildwood, the temperate jungle that carpeted Britain, was cleared in many places to create fields and pasture. The climate changed to favour the farmers as after *c* 4100 BC the summers grew warmer and encouraged the more widespread growth of cereals. Winters were colder but it rained less. But above all, men and women began to make an emphatic mark on the landscape. Amongst the most impressive monuments to the arrival of farming are now invisible. These were a series of mysterious wooden halls built by the incomers. All that remains now are the shadows of the postholes, but examples have been found in Aberdeenshire, Perthshire, on the banks of the River Tweed, in Oxfordshire and in Kent. It seems that all of them were burned down after a century or so of use, and archaeologists believe that they may have been the homes or even palaces of an elite. When these prehistoric lords died, the great timber halls probably became their pyres.

Farming not only triggered a population explosion, it also created the circumstances in which hierarchies could be built up. The key to this social process was agricultural surplus. Even more than hunting and gathering, farming is seasonal, it sets down a calendar that demands the completion of a series of tasks at specific turning points of the year. Over 60 centuries the rhythm of these tasks has changed little. Spring ploughing or ground preparation was done as soon as the frosts allowed so that the weather could help break up the soil even more and allow it to absorb the nourishing muck of the beasts which had been wintering on that ground. Prehistoric fields, like their successors in the Celtic west (cultivated and grazed well into the modern period) were small, and around a farmstead such as Knap of Howar they were used as enclosures for animals that were not slaughtered before the onset of winter. Their muck was an important means of fertilising fields that were in near-constant use.

Fields and Flocks

In a recent study, the distinguished archaeologist Francis Pryor brings his personal experience of farming into play and as a result has much to say that is sensible about prehistoric farming in Britain. One of his principal contentions is that the landscape was much better suited to stock rearing than the cultivation of cereals. The latter were grown only on a small scale, perhaps in garden-like enclosures. And because he understands their importance as a farmer, he makes much of the importance of markets in prehistory. Exchange took place and one of the principal reasons was the avoidance of in-breeding amongst sheep and cattle – and human beings. Pryor also believes that sheep were much more extensively reared than cows since they offer three crops: meat, milk and wool.

Spring ploughing and sowing could only begin after the ancient journey of transhumance. When the cold had at last come off the land and the green shoots began to push up through the bitter grass of winter, herdsmen moved their beasts up the hill trails to the high summer pasture. Some stayed on at the shielings to tend to cows, ewes and nanny goats as they dropped their young in the spring and their milk came through. This was the first renewed source of food for animals and people after the hungry months of the winter and it is celebrated in the old Celtic feast of Imbolc, a turning point still dimly recalled in the crofting communities of Scotland and Ireland.

In the months before the harvest and while their beasts were still grazing the upland pasture, communities had time for work not necessarily associated with food production. In good years a surplus will have allowed some to avoid the necessity of working on the land at all. In this way hierarchies began to stratify the society of prehistoric farming.

The great complex of monuments on Orkney and those of Wessex in the south stand testament to the skill of farmers in producing consistent surpluses. The millions of man-hours needed to build all of the henges, tombs, the strange linear ditches and banks known as cursus monuments and the enclosures of various kinds of the fourth millennium BC would simply not have been available if the new farms of Britain had not been tremendously productive, and the population not risen so sharply. Some of the calculations are spectacular. The stone circle at Avebury needed a million man-hours to make and the cursus in Dorset half a million.

The great monuments were also made possible by leadership of some kind, probably, in modern terms, a combination of the spiritual and secular. Many of these structures and layouts not only required the labour of many hands, they were also logistically complicated, especially in the transport of materials. Famously, the sarsen stones in the outer circle of Stonehenge were brought to the site from a considerable distance, probably a quarry that lay 25 miles away. The stones weighed 25 tons and *c* 2600 BC, their move-ment would have involved a huge-scale operation and the muscles of many people. The trilithons of the inner circle, the uprights and lintels were dragged along the same route and they were twice as heavy. Perhaps the most astonishing example of prehistoric haulage is the altar stone. Although it weighs only four tons, it was quarried in Wales, either in Carmarthenshire or the Brecon Beacons, and brought at least 80 miles to Stonehenge. And even more astonish-ing is the fact that 43 of these bluestones exist on the site, each weighing between 2 and 4 tons, and some archaeologists reckon there were originally 80 in all. New research suggests that these were brought some time around 3000 BC and were first used as grave markers. More than 50,000 fragments of cremated bone from the skeletons of 63 individuals have been analysed. New techniques have revealed that the burials were not exclusively of men but also included women and children, one of them a new-born baby. It seems that the early stones of Stonehenge were tombstones in a prehistoric graveyard, and perhaps one that held the remains of elite families rather than only elite men. How these massive objects were moved (to say nothing of why they were chosen) can only be a matter for conjecture, but what is beyond doubt is the organisation, communication and leadership needed to bring them to the great stone circle.

There must have been decision-makers, directing minds behind these projects, and efficient farming supported them. In addition to these shadowy leading figures, the priest-kings or -queens, there are likely to have been other specialists, perhaps members of their household, probably military. If the settlement of Britain and Ireland by groups of incoming farmers was accompanied by violence and the retention and protection of good land needed to be backed by force, then it is likely that the leaders of society, especially those who could command undertakings on the scale of Orkney and Wessex, were supported by warbands, soldiers who could develop the skills, strategies and weapons needed to enforce, if required, the dragging of the massive stones across Salisbury Plain so that they could be pulled upright at Stonehenge. Their identity will probably never become clear, but their DNA was very likely to have been that of the invading farmers, the carriers of the G lineages.

These markers announced the arrival of people whose DNA is closely linked – and this event is also mirrored in the archaeological record. Between *c* 3800 BC and *c* 2500 BC the land was cleared and the first monuments rose in the landscape, and what is striking is the clear cultural ties between them – and across very long distances. A distinctive motif which appears on pottery made in Orkney also turned up at a dig near Oxford and on a mace-head found in the Boyne Valley in Ireland. There are very obvious connections in the style of henge-building between monuments in the Thames Valley and in North Yorkshire. Cumbrian stone circles share close similarities with some in Ireland. Ideas, beliefs and techniques seemed to be moving freely across Britain and those ideas were, of course, carried in the minds of our prehistoric ancestors. It was clearly a dynamic, innovative and aggressive society capable of changing the landscape utterly and planting it with prehistoric cathedrals whose majesty is at least the equal of any structure raised since and, it may be argued, that they are greater achievements since they have outlasted many great buildings that have been raised since than, and still have the power to inspire awe.

5

The Coming of the Kings

✷

I N THE CORNER OF A field on the south-east fringes of the
small town of Amesbury in Wiltshire there was feverish activity
and it was matched by a growing pitch of excitement. Cars had
been arriving all day and as work went on long into the evening of
3 May 2002, their headlights were needed to allow it to continue
as the sun set over Salisbury Plain and darkness fell. By 2 a.m., the
exhausted team at last packed their gear. Everything that could have
been done in the short time available had been done. On what had
begun as a routine archaeological investigation, one of the richest,
most detailed and most suggestive prehistoric burial sites in western
Europe had been unexpectedly discovered.

In Britain, planning regulations usually insist that developers
who wish to build on green or brownfield sites must first have them
surveyed by archaeologists. Bloor Homes, Persimmon Homes
South Coast and QinetiQ had acquired land between a loop of
the River Avon and the edge of Amesbury for 2,500 houses and a
business park. Despite the fact that Stonehenge stood only three
miles to the west, there were, apparently, no major monuments on
the site and archaeologists expected that the most they might find
were traces of a Roman cemetery. But on the morning of Friday,
3 May, the last day of the survey, all thoughts of routine work
rapidly evaporated. In the corner of a field which would soon be
covered over by a new primary school, archaeologists came across
two ground features that looked different, nothing like Roman
graves. The slight depressions may well have been unremarkable,
perhaps caused by old blowdowns of large trees whose roots had

ripped out the soil. Nevertheless, the turf was removed and digging began.

Almost immediately, bones were found, and then early Bronze Age pottery came out, items the excavators recognised as the distinctively shaped vessels known as beakers. Work was abandoned on the rest of the large location to focus on this fascinating find and as archaeologists carefully trowelled and brushed aside the soil, unlooked-for treasures quickly came to light. By the early afternoon gold had been uncovered, the oldest gold objects yet discovered in Britain. In the corner of the burial, two gold hair clasps, beautifully wrought, had been lifted out and carefully cleaned. The archaeologists knew that in the third and second millennia BC in the south of England gold was extremely scarce, and its removal from circulation to be set in a burial denoted someone of elite status.

From Wessex Archaeology's headquarters in Salisbury, only 7 miles to the south, more archaeologists arrived to help and a find centre was prepared. Because the objects that came out of the grave were so precious, so opulent and important, and because the May Bank Holiday Monday to come meant a long weekend when the site could not be left unattended, the team worked on into the night until all was safely complete and the finds removed for safe-keeping back at base. What they had discovered was extraordinary, a series of objects and a skeleton that would shine an unexpected light on Britain in the third millennium BC, on what is known as the Bronze Age, a time when the coming of metal and metalworking would change society. And a time when Britain would see the coming of powerful, controlling figures: a time of kings.

Large and rectangular, the grave had probably been lined with timber shuttering and perhaps roofed over before being covered with a mound of earth, a round barrow. Before that happened, it is very likely that some sort of burial ceremony took place. Whatever his identity, the dead man was important and he enjoyed tremendously high status since his grave had more than 100 objects placed in it, ten times more than any other yet found. These were not only symbols of that high status but also likely to be items that might be needed in an afterlife, what are known as grave goods. And archaeologists were able to date the burial to *c* 2300 BC.

The corpse had been posed on its side in a crouched position and the places where some of the objects were found were very informative about how he had been dressed after death. He was an archer, or at least someone whose status demanded that he be laid to rest wearing archery gear and with archery equipment placed around his body. This conferred an attractive and alliterative identity and very quickly the man became known as the Amesbury Archer. His DNA marker remains to be discovered.

On his forearm he had worn a slate bracer or wristguard. Strapped to the arm that held his bow, it was designed to protect it from the recoil of the bowstring after he had loosed off an arrow. There was a second wristguard placed by the Archer's knees and from the location of 16 tanged and barbed flint arrowheads, it seems that a quiver of arrows had been strewn across his legs. Their wooden shafts and indeed the longbow itself had long since perished. More arrowheads, some flint blanks and other tools were found in a cache, which suggests that they had been contained in a pouch.

When archaeologists examined the skeleton itself, they found something that did not quite match the dashing image of a huntsman. Although the Archer was strongly built and had been healthy enough to live long for the times, dying between the ages of 35 and 45, he would not have been able to move easily and was, it seems, in constant pain. Perhaps in an accident, or perhaps as a result of a deliberate act to restrict his movement, the Archer's left kneecap had been ripped off. As a consequence he walked with a severe limp, with the injured leg probably held straight, perhaps swinging along as though it were an unarticulated wooden leg. He also suffered an acute infection in his bones which must have left him in chronic pain. If the Archer was a real and not a symbolic huntsman, he cannot have taken part in a chase of any sort or in the rigours of warfare. Rather, it is likely that he waited with his arrow notched on his bowstring for animals to be driven towards him in the ancient hunting method know as the drive and sett. Just as they do in modern pheasant or grouse shoots, beaters traditionally drove prey towards the sett where hunters with shotguns or bows and arrows could kill them as they flew or fled. Lying close to the Archer's skull was a find that suggested he was indeed a real huntsman: several tusks from wild boar, genuine trophies – if he killed them himself – for these animals were large and very dangerous when they charged out of cover.

In addition to the gold hair clasps, other useful objects were found and something that may explain the Archer's very high status. Three copper knives had been placed in the grave and behind the back of the skeleton lay what is known as a cushion stone, a small last or anvil for working metal objects. The inclusion of this item seems more than symbolic. Far more significant for the story of prehistoric Britain than the archery gear, the cushion stone talks of the arrival of new technology, of the coming of metalworking. The skills of a blacksmith like the Archer will have seemed magical, as dull ore was transformed by fire and with the use of mysterious means made into lustrous and beautiful gold and copper objects. Perhaps a better name for this remarkable man would have been the Amesbury Alchemist. His ability to make aesthetically pleasing pieces like the gold hair clasps and copper knives with a blade-edge honed to razor sharpness would not only have been highly prized but they may even have conferred a leadership role. When the grave was discovered in 2002, the more excitable newspapers reported that the King of Stonehenge had been found. But it is not an altogether fanciful projection – a unique monument that clearly involved the deployment of huge resource may have needed a king to direct and command. A very different and more thoughtful assessment of his importance to his contemporaries wondered if the Archer had been deliberately disabled so that he and his much prized alchemical talents could never leave the community around the great stone circle.

Much about the burial is a matter of inference, even pure speculation, but one vital piece of information that it gave up is incontrovertible. The Amesbury Archer was an immigrant. Tests on the enamel of his teeth showed that he grew up in central Europe, in the alpine regions of Austria, Germany and most likely Switzerland and came to Britain as a young man, bringing the new skills of metalworking with him. Did he come alone? Was he part of a group or a leader of incomers with revolutionary ideas? More clues were waiting under the grass at Amesbury.

Dating to the same period, another very similar burial was found close by. This time the man had died relatively young, between the ages of 25 and 35, and like the Archer he had been interred with two gold hair clasps wrought in the same style. And even more significantly, archaeologists discovered that the two men were closely related. In their feet, they shared a very unusual bone structure, something that had been passed down in their DNA. Their heel bones had a joint with one of the upper tarsal bones of the foot, an extremely rare phenomenon. They may even have been father and son: the King and Prince of Stonehenge? A similar analysis of tooth enamel was done but it yielded a different result. Even though they were related, unlike the Archer, the younger man was not born in alpine Europe but probably grew up in southern England and may also have spent time in the Midlands and north-east Scotland. Nevertheless, the fact of their burials being so close together does strongly suggest the coming of a group of immigrants from alpine Europe.

What made the beaker pots, sometimes called bell-beakers, different was their fineness, their thin walls and the characteristic decoration added by pressing twisted cords into the clay while it was still wet. Examples have been found in more than 2,000 sites in Britain and Ireland. But the earliest were made in what is modern Portugal, around the wide estuary of the River Tagus some time around 2800 BC. Used principally as drinking vessels and sometimes containers for food, beakers were associated with metalworking and hunting equipment from the very beginning. Copper, soft and relatively easily mined, was, with gold, the first metal to be worked and many small daggers or knives like those in the Archer's tomb were made. Discovered as grave goods, all of these characteristic items came to be known by archaeologists as the beaker package.

Before it reached Amesbury, this new culture spread north from Portugal by an Atlantic route and then fanned out widely across Europe within a very short time, between *c* 2700 BC and *c* 2600 BC. This was far too rapid for it to be a process of adoption; the movement of the beaker package must have involved the movement of people over long distances. Like the Amesbury Archer. DNA evidence makes a link right across Europe. At a beaker cemetery at Kromsdorf in central Germany, scientists have been able to extract viable DNA from six individuals. Two carried the mitochondrial markers I1 and K1. And two men from Kromsdorf carried R1b, a Y-DNA marker that is now found in very high frequencies in north-western Europe and has not yet been seen in any other skeletons. This is also well represented in Portugal and Spain. In the deeper past, R1b-M269 came from the Near East and it may have reached Germany by a more direct route. Alternatively it could have come to Portugal through the Mediterranean before men who carried it took the skills to create the beaker package north into Europe. The mtDNA markers, I1 and K1, are also originally from the Near East. What all of this patchy evidence suggests is a faded map of Europe, the traces of possible migration routes in and before the third millennium BC, and also that carriers of R1b were still arriving in Britain and Ireland 1,000 years after the coming of farming. What is striking is how the DNA evidence chimes with what archaeology has discovered about the beaker phenomenon and the introduction of metalworking. Taken together, the findings of these two disciplines sheds a steadier – even sensational – light on our prehistory.

A staggeringly high proportion of European men, most of them concentrated in the western regions of the continent, carry the marker labelled as R1b-M269. Geneticists reckon that 110 million have it and its sub-groups. The distribution of R1b-M269 is another clear example of gene-surfing, where a marker at the front of a new wave of migrants has a disproportionate demographic impact. In eastern Turkey only 12% of men carry it but in England it is 60%, in Scotland 72% and in Ireland an overwhelming 85%. This increase in frequency from east to west obviously shows the direction of travel and the marker's origins. R1b-M269 is the DNA of a second wave of farmers and as this revolutionary way of life, of producing food, washed across Europe, this marker, which first arose in the Near East, became tremendously widespread. And its dynamic movement to every corner of the continent shows something unarguable, that this new wave of farming and farmers was not a process of acculturation or adoption. Rather, it was the cultural cargo of an invasion, the deposit of a takeover by new people, the bearers of the marker of R1b-M269.

Recent research has dated the origin of the cluster of R1b Y-chromosome markers to the middle of the third millennium BC and attached its dramatic spread across Europe to the advance of the Beaker People. And it seems that there was conflict, perhaps even something close to genocide. When the carriers of the R1b lineages arrived in Britain and Ireland, they took over, and this process must have involved the elimination of many of the men who carried the

G lineages of the first farmers. The ability of the newcomers to work metal and make magical objects from gold and copper may have conferred prestige, but their military superiority was what mattered. Perhaps the key to that was also found in the Amesbury Archer's grave. It may be that the carriers of R1b were expert with bows and arrows and were able to seize land quickly by force of arms. Historical analogies are not unhelpful. During the Hundred Years War of the fourteenth and fifteenth centuries, English archery dominated the battlefields of Agincourt, Crécy and Poitiers, often defeating larger and much better armoured French forces. More finds from the third millennium BC continued this thread.

Across the A345, a very short distance from where the Amesbury Archer was found, another beaker burial came to light in 2003. Roadworks for the Boscombe Down airfield alerted archaeologists and they turned up the remains of seven individuals: three children, a teenager and three men. Because archery equipment, a boar's tusk, flint tools and eight beaker pots were also found, this group became known as the Boscombe Bowmen. Analyses of their tooth enamel suggested that they had come from Wales or the Lake District and the shared grave was dated to *c* 2300 BC, approximately the same time as that of the Archer and his young relative.

At least part of a picture begins to form when a fourth find is considered. In 1978, a skeleton that was dubbed the Stonehenge Archer was discovered in the outer ditch of the great stone circle. He was local, dated *c* 2300 BC, had a wristguard and there were flint arrowheads. But the tips of several were found in the bones of the skeleton and it is believed that he was killed by a group of archers. Was he killed by men who were not local? Perhaps he was a native leader who was ritually executed by the prehistoric equivalent of a firing squad before his body was placed in the sacred earth of the ditch.

The ability to make weapons from metal was almost certainly a second factor in the military superiority of the carriers of the R1b lineages. The razor-sharp copper daggers found in the Archer's grave could have been lethal in close-quarter fighting and a copper-bladed halberd discovered in Ireland with its wooden shaft intact was a fearsome weapon.

100

At the same time as all of these people were being interred in the landscape around Stonehenge, it was changing. The simple ditch and bank of *c* 3000 BC was made infinitely more dramatic 700 years later as the huge 20-ton trilithons were erected, having been quarried from the nearby Marlborough Downs, and the smaller bluestones were dragged 240 miles from Preseli in west Wales. What prompted and enabled this striking cultural change? Was it the arrival and triumph of the new farmers, the carriers of the R1b lineages? It is impossible to know for certain but the archers of Stonehenge arrived at approximately the same time and they definitely brought decisive cultural change. Was the Amesbury Archer a priest-king, a man who could conjure the magical metals of gold and copper into lustrous life with his alchemical skills? Was he part of a new, incoming elite who inspired and oversaw the creation of a new Stonehenge, one of the most singular structures ever raised? It seems too much of a cultural coincidence.

The archaeological consensus agrees that Britain's beaker people and their new way of life crossed the English Channel and the southern basin of the North Sea in an arc from Brittany to the Low Countries. This in turn implies two migration routes: one that traversed continental Europe from the east and another from the south, an Atlantic approach, probably sea-borne. It seems that people with skills that necessarily complemented those of the Amesbury Archer and other coppersmiths came to Britain from the latter direction.

The flashy pyrotechnics of metalworking needed the much more laborious and mundane efforts of mining. Copper is the world's eighth most common mineral and the deposits that existed in Britain and Ireland required to be opened up if more attractive objects such as the daggers of Amesbury were to be made. The earliest workings have been dated to *c* 2400 BC at Ross Island, near Killarney, but the most extensive prehistoric copper mine in the world was dug at Great Orme Head by Llandudno on the coast of north Wales. More than five miles of tunnels and passageways have been found and the first ore was extracted at about the same time as the bluestones were being dragged from Preseli in west Wales to Stonehenge. Over the first 1,000 years of the mine's life, about 50,000 tons of rock and ore was hacked out with antler picks and granite stone hammers. Great Orme was an industrial-scale operation.

Even though the work of the miners was less obviously skilled than that of a coppersmith, it required experience, strength and courage. When the first blows were struck on the headland, they were not struck by someone new to the job. At Great Orme, Ross Island and elsewhere, miners had to know exactly what they were looking for, the rocks that would contain copper. Experience was vital, since they worked in dark tunnels lit only by candles made from animal fat which could be blown out by draughts at any moment, and some of the passageways were so narrow that they could only have been excavated by small children, perhaps only five or six years old. Great Orme became a warren of tunnels which only stopped delving deeper down into the earth when the miners reached the water table.

Once ore-bearing rock had been extracted, it was broken up to separate the copper-bearing deposits from the gangue, the worthless residue. Here is a good description of the ancient process from Lewis Morris, the Crown Mineral Agent for Cardiganshire. He was writing in 1744:

Their method seems to be this. They make a great fire of wood in the bottom of their rakes [pits] which were always opened up on that account, and when the rock was sufficiently hot they cast water upon it, which shiver'd it, and then with stone wedges, which they drove in with other stones, they work'd their way through the hardest rocks, tho' but slowly.

Once the ore had been extracted and broken down into manageable lumps, it could then be smelted to produce ingots of copper, and that is what the Amesbury Archer will have worked on his cushion stone.

If the speed of the spread of the beaker package insists on the movement of peoples with the associated skills, then miners who knew what they were doing are very likely to have migrated to Ross Island and Great Orme. DNA offers a fascinating trace of such journeys.

The first evidence of metalworking and mining in Europe was found in the Balkans, the area that is now Serbia. A copper axe dating to *c* 5500 BC was found at Plocnik and the raw material needed to make it was extracted at Rudna Glava, about 100 miles to the north. The techniques and logistics of mining and metalworking were clearly well established in the Balkans and were later transmitted westwards. DNA analysis shows how far west these skills travelled.

The Y-DNA marker E-V13 originated in the Balkans where it is to be found at its highest frequency, 39% of all men in the region of southern Serbia and Kosovo. Frequencies of 8% in southern Italy and 7% in Sicily, areas colonised by Greeks towards the close of the first millennium BC, show how many carried the marker westwards. In Portugal, E-V13 accounts for 3% of all men. But what is very striking is its appearance in north Wales, very close to the mines at Great Orme Head. In a survey of men living in the town of Abergele, about 4 miles east of the mines, a very surprising 38% carried the Balkan marker. They may well be the descendants of the first miners to hack at the rock in the tunnels under the headland in search of copper, what appears to have been the raw material of power in Britain at the end of the third millennium BC.

As a soft metal, easily dented and bent out of shape, copper and the objects made from it had limited practical use. It was almost certainly valued principally for its rich colour and lustre, much like gold, and as such it was at first closely associated with elites. And these must have been powerful people, for the effort and organisation required to produce objects made from copper were immense. Perhaps they also had a religious quality. Other cultures appear to have believed that metal of any sort had an immanent magic. Around 3000 BC Egyptian weapons made from meteoric iron were known as Daggers from Heaven.

Clearly in what is sometimes called the Copper Age in Britain there were leaders, perhaps kings and their retinues, who could command and enforce the huge efforts needed to mine and smith objects of little practical value. The fact that the coming of copper- and metalworking to Britain appears to have had no quotidian impact on the lives of ordinary farmers is perhaps less important than what it had to say about society generally. The appearance of these new objects unequivocally confirms the formation of a hierarchy. At the summit of the pyramid sat kings, or people with an equivalent status, while below them specialists worked to supply all sorts of perceived needs: the warriors, the miners, the smiths, the quarrymen, those who transported goods, the builders of monuments and those who animated them spiritually in whatever unknowable way, and then supporting all of them, the vast majority – the farmers and herdsmen of prehistoric Britain who toiled to produce food and the necessary surpluses.

Archaeologists rightly make much of the accident of survival, of how our view of prehistory (or indeed all history) should not be conditioned by the extraordinary monuments in Orkney and Wessex and the fact that there is comparatively little to be seen above ground between them. And yet instinct insists that these two centres were indeed focuses, a factor that may have made their survival a little less of an accident. Stonehenge stands in a wide and detailed ritual landscape with spectacular circles at Avebury, Durrington Walls, Woodhenge and elsewhere. Many other upstanding features can be added, such as Silbury Hill, and others less obvious like the Dorset Cursus. Nevertheless, it is the magnificent circle of trilithons that appears to act as a magnet – and perhaps it has been seen that way for more than 4,000 years. They could be the most singular monument to the arrival of the new farmers, the carriers of the immensely widespread cluster of R1b markers.

In Europe the spread of the Beaker culture can be seen not only in Y-chromosome DNA. Recent research has shown that mtDNA also saw dramatic changes in the third millennium BC. DNA sequenced from around 40 ancient skeletons in central Europe showed dynamic change. These carried the mtDNA marker of H. Along with its sub-types it is the most common in Europe at a frequency of 40% to 50%. But in the time of the early farmers, the first wave, its frequency was about 19%. And then it expanded very rapidly, with the creation of new sub-types. Geneticists also found close similarities with people who carry H in modern Spain and Portugal. Professor Alan Cooper of the University of Adelaide has co-authored a study about this which points out that, 'What is intriguing is that the genetic markers of this first pan-European culture, which was clearly very successful, were then suddenly replaced around 4,500 years ago.' It appears that women accompanied the waves of R1b farmers as they eliminated the carriers of the Y-chromosome G markers.

It seems likely that this invasion was cultural as well as military.

Oriental Jones amused himself by using the pen name of Youns Uksfardi, or Jones of Oxford. Perhaps any variations of plain William Jones were to be welcomed. But in truth William Jones was far from plain. Born to Welsh-speaking parents in London in 1746, he quickly impressed them with his precocious fascination for and facility with language. William Jones Snr and Ann Jones were both from Ynys Mon, or Anglesey, and bilingual, but by the time their brilliant son was barely into his teens he had also become fluent in Greek, Latin, Persian, Arabic, Hebrew and could write in Chinese. On his early death at 48, this extraordinary man had complete command of 13 languages and could get by in another 28. A hyperpolyglot, Jones carried entire lexica in his head, swarms of words swirled around and could be instantly retrieved and marshalled into fluent expression. As he read, wrote and translated, what began to form in his prodigious mind was an understanding of how it was that one individual could understand so many of the world's languages. Jones could hear them talking to each other in his head. He realised that many of them had a great deal in common.

In 1783 William Jones was knighted, appointed a judge on the Supreme Court of Bengal and sent out to dispense justice in the vast new British colony of India. It was to prove a transformative experience. Jones's duties on the Bengali bench cannot have been too onerous for he quickly found time to immerse himself in and become entranced by Indian culture. Learning several native languages, he found himself fascinated by Sanskrit, the classical liturgical language of both Hinduism and Buddhism. Like Latin and Greek for Europeans, it spoke of great antiquity to the Indians of the late eighteenth century – and it spoke of something even more profound to William Jones.

Counting in Indo-European

English	Old German	Latin	Greek	Sanskrit
One	Ains	Unus	Heis	Ekas
Two	Twai	Duo	Duo	Dva
Three	Thrija	Tres	Treis	Tryas
Four	Fidwor	Quattuor	Tettares	Catvaras
Five	Fimf	Quinque	Pente	Panca
Six	Saihs	Sex	Heks	Sat
Seven	Sibum	Septum	Hepta	Sapta
Eight	Ahtau	Octo	Okto	Asta
Nine	Niun	Novem	Ennea	Nava
Ten	Taihum	Decem	Deka	Dasa

The hyperpolyglot noticed that all three classical languages appeared to share structure and vocabulary. The words for counting seemed particularly close. For example, three was almost identical: trayas in Sanskrit, tres in Latin, treis in Greek seemed to be the ancestors of tres in Spanish, tre in Danish, Swedish and Italian, trois in French, drei in German, tri in Russian, Scots Gaelic and Welsh and of course three in English. The words for key family relationships are also clearly common with mother/moeder/mater and father/pater/patir as well as many variants. Here is what William Jones concluded:

The Sanscrit language, whatever be its antiquity, is of a wonderful structure; more perfect than the Greek, more copious than the Latin, and more exquisitely refined than either, yet bearing to both of them a stronger affinity, both in the roots of verbs and the form of grammar, than could possibly have been produced by accident; so strong indeed, that no philologer could examine them all three, without believing them to have sprung from some common source, which, perhaps, no longer exists; there is a similar reason, though not quite so forcible, for supposing that both the Gothic and the Celtic, though blended with a very different idiom, had the same origin with the Sanscrit; and the old Persian might be added to the same family.

William Jones died relatively young, but his pioneering aperçu was developed by an unlikely disciple. Jacob Grimm, one of the famous Brothers Grimm, became very interested in the history of the German language and saw it as an important branch of the Indo-European family. His approach to research was straightforward. Jacob set out for country districts and on his travels asked local people about rare words that may have survived in their vocabulary unchanged over time. When his questions were greeted with suspicion, Grimm was forced to change his approach. Instead he invited country people to tell him traditional stories and he noted down unusual words that they used. With his brother, Wilhelm, he later published them as Grimm's Fairy Tales.

And grim they sometimes were. What Jacob discovered through these old and often dark tales was a pattern of evolution. As Germanic languages developed out of early Indo-European, the clear lineage of vocabulary was obscured by what Grimm called the Great Consonant Shift. By analysing a series of transitions, it was possible to work back from a word like 'kin' to the Latin 'gens' for race or family, or from 'Vater' for 'father' to the Latin 'pater'.

The Grimms' successors also noticed that many family-based words were linked and could be traced back to what was posited as a proto Indo-European language. Some were intriguingly specialised. A set of terms for in-laws appeared to refer to the relatives only of a bride and not a groom. This in turn implied a patriarchal society of long standing into which a bride was accepted and her relatives were as a result more closely defined. The groom's family were simply 'the family', the core social unit. Now, this is a slender thread upon which to hang a theory, but when other common linguistic connections are brought into the argument, the case for the existence of a male-dominated society in at least part of our prehistory is strengthened. The root word in several major Indo-European languages for king is 'reg', used as 'rex' in Latin and surviving in English as 'regal'. In the older Celtic languages such as Scots Gaelic, a queen is a 'ban-righ', literally a woman-king.

Many more commonalities are suggestive of the culture of early speakers of Indo-European. For example, there is a repeating emphasis on triplicity, the habit of thinking in threes. This ranges from the spiritual (the Father, Son and Holy Ghost) to a triple-tiered structure for a society of ruler-priests, warriors and herdsmen-farmers. The latter is seen from ancient India through Iran to Greece, Rome and what Julius Caesar wrote about Celtic France or Gaul (which was famously divided into three parts). Also shared are various versions of the notion of an over-arching deity seen as a Sky-Father, and both the Roman Jupiter and the Greek Zeus were thought of in that way.

Linguistic coincidences begin to multiply into likelihoods when the anthropological nostrum that a common language almost always presupposes a common culture is borne in mind. And much academic ink has been spilt, and even some academic blood, on where the original Indo-European-speaking culture arose and why it spread. A growing consensus agrees that it began in the Fertile Crescent, the arc from modern Iraq through Syria to the eastern Mediterranean shore – and that it was adopted beyond this region with the spread of farming.

The argument runs that as new techniques of food production and a new way of life were imposed on a native hunter-gatherer population (even if this meant only the women of that population as they were taken by incoming men as partners) and that package came described by a new language, it swept – almost – all before it. Perhaps it is no coincidence that the ancient range of wheat-growing (as opposed to its modern extent in the Americas, China and elsewhere), from India across western Asia and Europe, exactly mirrors the extent of the Indo-European language family, with only a handful of exceptions. And much of the terminology specific to farming is shared across that area: grain/granas/gran and the words relating to cows, sheep, pigs, horses and much else are clearly cognate.

The European languages that stand outside the Indo-European family are Basque, Finnish, Estonian and Hungarian. In each of these enclaves, large communities may have resisted the adoption of new ways to describe a new life. In the case of Basque, the suspicion is that it may be an even older language; perhaps its very different basic vocabulary echoed around the painted caves of the ice age refuges. And without stretching conjecture to breaking point, it may have resembled the language spoken by the women of Britain, the pioneers who came back after the ice, the carriers of markers H and V and their sub-groups when they encountered the first waves of the first farmers, the men enriched for G.

What is certain is that the oldest living languages in Britain are to be found in the west. Scots Gaelic and its Irish and Manx cousins, and Welsh and Cornish are still spoken and inside them survives a sense of a much older way of understanding and describing Britain, one that may be closer to how the first farmers saw the world. As members of the Indo-European family of languages, they speak of a life on the land, subsistence farming, the day-in, day-out labour of the vast majority of a pre-urban population. They talk of an intimate and highly detailed understanding of the weather and all its moods, of the habits and appearance of animals and of the landscapes and seascapes their speakers depended on.

Far from being relics of a romantic Celtic past, Gaelic and Welsh are as lexically tight and as precisely descriptive as Ciceronian Latin. They had to be. For farmers, the weather governed everything and Scots Gaelic, for example, often looked to its western horizons, to the cloud formations over the Atlantic and their effect on the ocean. English no longer has need of a word like 'stuadhach' but farmers on the shores of the Hebrides did. It describes a day when the breakers are rolling in off the ocean and are 'billowy, surgy, huge waves with steep white peaks'. Precision of this sort mattered to those who worked on or near the sea. Words like calm or stormy were simply too loose to be of much use.

Richness and detail decorate the pages of the lexica of the Celtic languages of Britain, but for the overwhelmingly urban population of the twenty-first century, they reflect a world that is fast submerging. Scots Gaelic in particular may be a last fraying link with the early farmers of prehistory.

Memories of the beginnings of Britain, the return of the pioneers after the ice and the coming of the farmers, will have depended absolutely on an oral tradition, something that by definition was nowhere recorded in a fixed form, tales that have been long forgotten or remade into something so different as to lose almost all sense of their original narrative. Except perhaps in some unlikely places.

Like many of the peoples and places of Europe, Britain and its nations have a collection of origin legends, stories of founders and first arrivals. Myth-historical, whimsical and often political, they are not to be relied upon as a historical record. But neither should they be ignored. One of the earliest is arguably the most informative, the *Lebor Gabála Érenn*, the 'Book of the Taking of Ireland'. It first came to light as an eleventh-century manuscript that recorded much older stories of the peopling of the island. Four invasions are said to have washed on Ireland's shores and they may be faint echoes of real immigrations, for their compass direction agrees both with what DNA can tell and what archaeologists have found. The *Lebor* recounts the coming of people from Iberia (even allowing for scribal confusion with Hibernia) by an Atlantic route and it asserts that they were Gaels, that is, speakers of early versions of the Gaelic language.

Nennius, or Nynia, was probably an eighth-century monk from north Wales who compiled a *History of the Britons*, engagingly confessing that 'I have made a heap of all that I have found'. In setting down the stories of the natives of Britain, those whose kings had been pushed west by the invading Anglo-Saxons, he invented or borrowed a great deal that is clearly mythic and much that is intriguing, such as the earliest notice of the campaigns of a historical Arthur, whom he calls not a king but the Dux Bellorum, the General. And again he brings the migrants to Britain from credible directions, from the western Mediterranean, and he repeats the stories of the incursion of people from Spain to Ireland. Nennius also created a founder-figure, Britto, and notes that he crossed from Gaul or France to Britain, which was promptly named after him.

Perhaps the greatest historian before the modern era, a truly rigorous scholar, was Bede of Jarrow. Even though it seems that he was born on the lands of the twin monasteries of Monkwearmouth and Jarrow, did not stray far during his productive life and relied on a library that seemed vast to him (but would seem small to us), letters from informants and other fragments, Bede wrote a superb and highly political account, *The Ecclesiastical History of the English People*. There are no giants, no Trojans and no Britto, but he asserted that four nations lived on the island of Britain: the English (with whom he is principally concerned), the Scots, the Picts and the British. Writing in the early eighth century, Bede chronicled the coming of the Angles, the Saxons, the Jutes and other Germanic kindreds but even though he says little about them, he understood that the British were the original inhabitants. 'According to tradition', they came by an Atlantic route, from Armorica or Brittany. This is probably a confusion since in the fifth century migrants sailed in the opposite direction to give Brittany, or Little Britain, its name. Nevertheless the general trend and the cultural connections are correct. Bede also reckons that the Picts crossed to the Scottish mainland from Scythia, by which he probably meant north-western Europe. This may well be a memory of the coming of farmers, those who made the pottery known as Linearbandkeramik.

The most famous – and notorious – chronicler of the beginnings of Britain was an Oxford cleric in the twelfth century, who came from Wales and probably spoke Welsh. In his sprawling account, Geoffrey of Monmouth created the popular medieval version of the tales of King Arthur, complete with his knightly companions, Excalibur, and Merlin but his own very real cultural milieu encouraged the Welshman to remember that the original British, those whom Arthur sought to protect, were a Celtic people and that England had once been a Celtic nation. Scholars who followed Geoffrey, Nennius and Bede repeated much of what they had written and occasionally, like the schoolmaster William Camden, added significant elements. In his great history, *Britannia*, published in the early seventeenth century, Camden asserted that the first of the British came from Gaul. And, displaying considerable acuity, he noted that the name 'British' meant 'The Painted People'. The late seventeenth-century curator of the Ashmolean Museum in Oxford, Edward Lhuyd, amplified the Spanish origins of the British and Irish Celts.

Whatever the weaknesses, strengths and vagaries of all that

scholarship over 1000 years and more, many of these chroniclers appear to echo memories, their tales linked to real migrations, to genuine places of origin or at least transit. But it seems highly unlikely that these murmurings of an unrecorded, half-forgotten deep past could refer to the slow and sparse process of repopulation after the melting of the ice. Instead they probably recall the dramatic impact made on Britain over a much more concentrated period by the coming of the farmers, the men who carried the marker of R1b and their seizure of the land.

Medieval monasteries used to admit laymen to become novice monks towards the end of their lives, often in return for an endowment or gift of some sort. This was known as taking holy orders *ad succurundum*, or in a hurry, and the point of it was to be sure of a cleansing burial. In the Middle Ages, the grounds or precinct of an abbey, convent or great church were believed to be holy, in a literal sense, because of the holy men and saints who had walked on it. More, the soil was thought to have the effect of cleansing a dead body of mortal sin and reducing the time spent in purgatory. The nearer the altar a body was buried the more powerful would be the effect. A similar way of understanding death and the passage to the afterlife may have been at work at Stonehenge at the end of the third millennium BC, a time when the culture of the Beaker folk seems to have been changing.

On Normanton Down, less than a mile to the south of the stone circle of Stonehenge, there is a very large cemetery of barrows, earthen mounds of various shapes and sizes that cover graves. Three are long barrows and nearly forty are smaller, round barrows. One of the latter was excavated in 1808 by William Cunnington and his patron, Sir Richard Colt Hoare. Even richer than the grave of the Amesbury Archer, it marked the introduction of a new phase in Britain's prehistory. This is an extract from the report of their excavation on what was called the Bush Barrow:

> On reaching the floor of the barrow, we discovered the skeleton of a stout and tall man lying from south to north: the extreme length of his thigh bone was 20 inches. About 18 inches south of the head we found several brass [bronze] rivets intermixed with

wood and some thin bits of brass nearly decomposed. These articles covered a space of 12 inches or more: it is probable therefore that they were the mouldered remains of a shield. Near the right arm was a large dagger of brass and a spearhead of the same material, full 13 inches long, and the largest we have ever found. Immediately over the breast of the skeleton was a large plate of gold, in the form of a lozenge, diminishing gradually towards the centre. We next discovered, on the right side of the skeleton, a very curious perforated stone, some wrought articles of bone, many small rings of the same material and another lozenge of gold. As this stone bears no marks of wear or attrition, I can hardly consider it to have been used as a domestic implement, and from the circumstances of it being composed of a mass of seaworms or little serpents, I think we may not be too fanciful in considering it an article of consequence.

They were not too fanciful. Cunnington and Colt Hoare had discovered a mace carved from sea fossils, an enduring symbol of great power, part of what one respected historian called Britain's first crown jewels. In addition to the two gold plates, which appear to share close stylistic links with similar objects found across the Channel in Brittany, there were three bronze daggers, a bronze axe head and bronze rivets. The hilt of one of the daggers was decorated with tremendously intricate and fine work, something not seen before in Britain. Set in a herringbone pattern, no fewer than 140,000 tiny gold pins had been placed in individually drilled holes. Each was no more than a millimetre in length, as fine as a human hair, and all had been glued in position before being polished to a brilliant lustre. It was an extraordinary example of sophisticated craftsmanship, done without the aid of magnification or precise metal instruments, a piece of delicate work that must have taken months, if not years. These very beautiful crown jewels testify to a powerful man, surely someone of royal authority.

'Tall and stout' was William Cunnington's description and the man in the grave is likely to have stood a head taller than his contemporaries, a giant at six foot or more, something that would have mattered, and been an asset for a warrior. In high contrast with the limping Amesbury Archer, whose metalworking skills appear to have been the reason for his elite status, the man in the Bush Barrow was surrounded by weapons of war, daggers, an axe and a mace. He was not a huntsman. There were no arrowheads or boars' tusks and no object like the cushion stone to hint that he did anything other than command. There were also no beakers and the date of the burial at *c* 2000 BC seems to signal the beginning of a fresh cultural shift. His weapons were forged from bronze, a new alloy that was probably perfected in Britain.

North of Helston in Cornwall, in the narrow valley of the River Cober, the Poldark Tin Mine stands astride 3000 years of history. Unlike copper, tin is a rare metal. Only in Cornwall and Devon are British deposits to be found and across the rest of western Europe supply is restricted to small-scale extraction in the Iberian Peninsula. These accidents of geology quickly became determinant since by *c* 2200 BC tin was vital to the prehistoric economy. Even though bronze objects were at first made only for an elite, large numbers of people became involved in their production. Made from around 88 parts copper and 12 parts tin, bronze was much harder than pure copper but retained its attractive sheen. It is also harder than wrought iron and, much later, officers in the Roman army carried bronze swords while ordinary legionaries and auxiliaries wielded iron blades. And unlike copper, bronze weaponry such as the Bush Barrow daggers was sharp, heavy and intimidating. With the key ingredient of tin in relatively plentiful supply and close at hand, it is likely that bronze-smithing developed first in Britain.

At Poldark, there existed no prehistoric mine. Tin was first found in streams or the beds of rivers like the Cober. Where lodes of the mineral were exposed on hillsides or cliffs, erosion caused the raw metal, known as cassiterite, to break off and lie on valley slopes as tin-rich gravel or pebbles. Rain eventually washed it down into streams and rivers as alluvial deposits. Sometimes these were very substantial, several metres deep and wide. Tin is easily recognised as dark and heavy pebbles or gravels. But tin-streams were not always found underwater and a version of open-cast mining was practised in areas where peat and earth had covered deposits. Over many upland parts of Cornwall and Devon, and especially on Dartmoor

114

and Bodmin Moor, these early workings are easily recognised by the mounds of spoil or 'overburden' that were laboriously removed before the miners could get at the tin.

Lodes and Shodes

Tin-mining in Cornwall has left its marks everywhere. In addition to the beautiful pit-head buildings that sometimes perch perilously on cliffsides and clifftops, there are many other telltale signs. Where lodes of tin were found at coastal sites, slot-like cuttings can still be seen. At Mount Hermon, near St Just, there is a large and very obvious nick that has been hacked into the tin-bearing rock. Shode workings are sometimes seen on hill-tops and plateaus. Deposits of shode or tin-rich stones that had been exposed by the erosion of an outcrop were worked and in some places the landscape is pock-marked by scores of shallow pits and soil dumps. The hillocks left by the tinners were called shambles.

Once these underground tin-streams had been exposed, the ore was panned through water rather in the same way as gold was in more modern times. When miners had been forced to strip off a great deal of overburden, they sometimes dug channels to divert water-courses to wash out the earth and other residue. This process was almost certainly witnessed by Pytheas, a Greek traveller from Massalia or modern Marseilles, who visited Britain *c* 325 BC. His own account of his extraordinary voyage, *On the Ocean*, has been lost but it was plagiarised by several subsequent Roman historians. Here is Diodorus Siculus, a contemporary of the Emperor Augustus, with an unattributed quote from Pytheas on his visit to the tin-streamers of prehistoric Cornwall, or as he called it, Belerion:

The inhabitants of Britain who live on the promontory called Belerion are especially friendly to strangers and have adopted a civilised way of life because of their interaction with traders and other people. It is they who work the tin, treating the layer which contains it in an ingenious way. This layer, being like rock,

115

contains earthy seams and in them the workers quarry the ore which they then melt down to clean it from its impurities. Then they work the tin into pieces the size of knuckle-bones and convey it to an island which lies off Britain, called Ictis; for at the ebb-tide the space between this island and the mainland becomes dry and they can take the tin in large quantities over to the island on their wagons. (And a peculiar thing happens in the case of the neighbouring islands which lie between Europe and Britain, for at flood-tide the passage between them and the mainland runs full and they have the appearance of islands, but at ebb-tide the sea recedes and leaves dry a large space and at that time they look like peninsulas.) On the island of Ictis the merchants buy the tin from the natives and carry it from there across the Straits of Galatia [the English Channel] and finally, making their way on foot though Gaul for some thirty days, they bring the goods on horseback to the mouth of the Rhone.

The bronze-making forges of the Mediterranean were hungry for British tin and the trade route described by Pytheas was no doubt busy. As techniques improved and demand surged, Cornish and Devonian miners began to dig into the lodes of tin ore, the sources of the deposits of the tin-streams. At Poldark, shafts were sunk into the hillside, into what became known as the rich Wheal Roots Lode, but on exposed cliffs such as Mount Hermon near St Just, cuttings can clearly be seen. It is as though a stratum of rock has simply been sliced out and a gap left.

The earliest recorded name for Britain was coined by the Father of History, Herodotus. Writing in the fifth century BC, his major theme was the war between the Persians and the Greeks from 479 BC to 470 BC, but his general interests ranged much wider. A scrupulous scholar, Herodotus was determined to write of what only he himself knew or what others had witnessed.When he came to deal with what he called 'the extreme tracts of Europe towards the west', Herodotus constantly qualified his assertions:

I cannot speak with any certainty for I do not allow that there is any river, to which the barbarians give the name of Eridanus, emptying into the northern Sea where, as the tale goes, amber is produced, nor do I know of any islands called Cassiterides, whence the tin comes which we use ... Though I have taken vast pains I have never been able to get an assurance from an eye-witness that there is any sea on the further side of Europe. Nevertheless, tin and amber do certainly come to us from the ends of the Earth. The northern parts of Europe are very much richer in gold than any other region: but how it is procured I have no certain knowledge. The story runs, that the one-eyed Arimaspi purloin it from the griffins, but here too I am incredulous.

Cassiterides means the Tin Islands and the logistics of the tin trade, which appears to have been well established by the time Herodotus was writing, are hinted at in the previous passage from Pytheas via Diodorus Siculus. While it is likely that ingots of smelted tin passed through several pairs of mercantile hands before they reached the forges of the Mediterranean, the location of the export market called Ictis seems clear enough. Off Marazion on the south Cornish coast lies the tidal island of St Michael's Mount. At low tide an ancient causeway between it and the mainland is exposed and at high tide the waters of the Channel rush in to make the mount an island once more, just as the passage from Pytheas says. It is very close to some of the most productive tin-streaming areas, including the valley of the River Cober. Beach markets were a common commercial phenomenon for thousands of years around Britain. Merchants with goods to trade were in the habit of beaching their boats at low tide to do business and then sailing off with local goods once they had been refloated at high tide. This way of making exchanges obviated the need for harbours and was suitable for small craft.

Between Marazion and St Michael's Mount it may well be that ships from Brittany came to trade for tin. Having heaved aboard the heavy ingots, they then recrossed the Channel or perhaps made their way north-west up the English coast before crossing to the mouth of the Seine and making their way inland. Either way, once the European mainland was gained and the passage afforded by navigable rivers exhausted, the tin was loaded onto packhorses and carried overland to the Rhône and then taken downstream to the Mediterranean. In the Cornish language Marazion is rendered as Marghasbighean. It means marketplace. And very intriguingly, traces of an ancient bronze-making forge have been found just outside the modern town.

Ted Wright and his brother Will were sharp-eyed amateur archae-ologists. Since the early 1930s, they had often gone out walking on the foreshore at North Ferriby on the estuary of the River Humber, only a mile or so east of the city of Kingston upon Hull. On the clay and peaty bed of the river, exposed at low tide, the brothers had occasionally spotted bones and antlers, and once a piece of blackened wood that had clearly been worked. All had been preserved in the anaerobic conditions. But in 1937 Ted found something remarkable. The powerful Humber tides had shifted a layer of the foreshore and Wright saw the ends of three substantial oak planks sticking out of the mud. Instantly, he knew that these were the remains of an ancient wooden boat.

After careful excavation over two seasons, the planks were recog-nised as the bottom of a long, narrow craft and one end of it had survived almost entire. What remained measured 13.1 metres long and 1.7 metres wide, and the complete original was probably 15.1 metres by 2.6 metres. Experts reckoned that when propelled by 18 paddlers, the Ferriby boat could have reached speeds of 6 knots, and with fewer on board, she could have carried between 3 and 5 tons of cargo.

While on leave from the army in 1940, Ted Wright found fragments of another plank boat at North Ferriby only 55 metres from the first, and in 1963 a third was discovered nearby. Perhaps the site had been a boatyard or a ferry depot. Carbon dating has put the construction between 2030 BC and 1680 BC, the first phase of the Bronze Age.

Since the Ferriby boats appear to have had a low freeboard, that is, only 40 cm between the waterline and the top of the side strakes, it is very likely that they were used in the estuary to take cargo and people across. In a rough sea, these sleek little craft would have been easily swamped. But on calm days, boats built in the second millen-nium BC probably plied a trade in inshore coastal waters. In 1992 improvements to the A20 in Dover led to the discovery of another one. Provisionally dated to *c* 1300 BC, it had scratches on the bottom from having been hauled up on pebble beaches, and pieces of stone from Dorset were found amongst its timbers. It looks as though the Dover boat moved goods and people along the Channel coast.

Until the modern era, the last two centuries, water was the fastest and safest means of transport, and also the only feasible way of moving goods in bulk. A spectacular find in 2011 showed how interlinked prehistoric water traffic could be in Britain. East of Peterborough, along the old course of the River Nene as it meandered into the Fens, eight logboats were discovered buried deep in a commercial clay quarry. Made from hollowed-out oak tree trunks, they were large, with the longest measuring 8.3 metres. And in the anaerobic conditions, they were very well preserved. Carving was still visible on their hulls and other finds nearby confirmed a vigorous Fenland community. There were bronze spears and swords, and even clothing had survived in the damp clay, silt and peat. It may well be that these narrow craft were fishing boats because eel-traps made from willow withies were found and their shallow draught will have been well suited to the waterways of the Fen. Perhaps they traded their catch of fish and eels with others and used their boats to take it fresh and quickly to different markets.

Boat Cemetery

It appears that the boats were deliberately sunk in a mysterious act of ritual deposition. The transoms used to make the stern watertight had been carefully removed, exposing the grooves into which they would normally have been slotted. The Bronze Age fleet was found near a place where metalwork – spear heads and other objects – had been deposited and the site lies only two miles from the prehistoric complex at Flag Fen. One boat was richly carved, both inside and out, another had fitted handles for lifting it out of the water and a third showed signs of fires having been lit on its wide, flat deck. The fish and eels caught in the fens would have been roasted and smoked on board. The boats could have been sunk in order to keep the timber waterlogged and prevent it from drying out and splitting. But they were never retrieved and that strongly suggests a sacrifice of some sort. What is certain is that the fishermen of the Fens were not farmers, and maintained the old life of hunting and gathering.

The overall impression of the early Bronze Age is one of mobility. In addition to the plank boats of Ferriby, Dover and elsewhere, logboats appear to have been used the length of eastern Britain (an equally ancient example was found at the Friarton Bridge at the head of the Tay estuary near Perth), while curraghs and coracles still glided on top of the waters and waves of the west. Bronze continued to be used mainly for the manufacture of prestige items such as weapons, and ordinary people probably had very few if any bronze or metal implements. None of the ancient boats had any metal fastenings, no nails or cleats, and all of the planks had been sewn together with withies or yew roots. It was the demand for bronze amongst the elite that affected people, that made mining communities in Cornwall and Devon and also prompted the trading networks to expand and become more dynamic. Greater mobility and the growth of exchange systems appear to have had an impact on the distribution of Britain's Y-chromosome DNA markers.

By the middle of the second millennium BC, there is a sense of two zones of seaborne trade, linked but distinct. In the west, the Atlantic routes from Iberia, Biscay and Brittany to the Irish Sea and as far as the Hebrides see a much more emphatic presence of the sub-group of R1b that has been labelled S145. It probably originated in Iberia and south-west France but it is found at a very high frequency in the modern population of Ireland at 67% of all men and at 45% in Scotland. By contrast, in England it falls to 20%. This was probably the consequence of continuing contact over a long period, not of a concentrated phase of migration as happened when the farmers came to Britain and Ireland in the centuries either side of 4000 BC and then again with the arrival of the Beaker people.

A different sub-group of R1b, S21, skews eastward. Only 6% of Irish men carry it, while 21% of Englishmen do. S21 is present in 13% of Scots, but these men are to be found mainly along eastern coasts. For example, the frequency of the sub-group is very high in Moray and Aberdeenshire ,where that of S145 is much lower. A similar pattern is repeated down the east coasts of Britain. Across the North Sea, in northern Germany, the Low Countries and in Friesland in particular, S21 is very significantly present. Again, this bias is likely not the result of an event but of a process of trade and exchange.

By the middle of the second millennium BC the weather seems to have been generally benign and the population of Bronze Age Britain had begun to rise. Up to that time agriculture appears to have been principally pastoral as the system of transhumance developed. Cereals were certainly grown but there is little archaeological evidence that the cultivation of wheat and barley was anything more than small-scale. But after *c* 1500 BC, land management began to change. Small fields were enclosed in lowland areas while even upland pasture was divided into individual holdings. On Dartmoor communities built an extensive network of stone banks called reeves, and these boundaries seem to have been laid out over a short period and in a planned manner. All these are signs of increasing population pressure.

Individual farms became recognisable units where some form of tenancy or ownership must have been in force. And farmhouses were also built. Around *c* 1500 BC the characteristic British roundhouse began to rise in the landscape.

From Kilphedir in Sutherland to Hampshire in the south, these ancient and ingeniously designed buildings have left their distinctive mark on the ground. All over Britain the Ordnance Survey marks 'hut circles', usually on higher ground where the footprints of roundhouses have lain undisturbed by ploughing or planting for 3,000 years. Their diameters often told of large structures of more than 10 metres across, and the resources needed to erect such a big house were very substantial. At Kilphedir archaeologists reckoned that one of the larger houses needed 650 mature trees for its infrastructure.

In essence these buildings developed out of the timber tipi design first seen at Howick in Northumberland, and they were basically conical in shape. In some of the earliest examples, the roof trusses were planted directly into the ground, chocked into postholes. Walls varied in height and the ends of the roof timbers were placed on the wall-heads. More support was supplied by an interior ring of load-bearing posts and in later and larger roundhouses this held up a first floor which could be reached by a ladder. Roofing could be thatch, brackens or turf.

The ring of internal posts made it logical to create a radial layout, like the spokes of a wheel, with compartments that could be screened off for sleeping or storage. In the centre was the downhearth, a fireplace ringed by a kerb of flat stones upon which cooking pots could be set. A variation was to project a larger flat stone into the hearth so that whatever potage that was bubbling in a pottery vessel could bubble faster. In the roundhouses of the elite, bronze spits and swees for cauldrons were set up and smiths produced a canteen of high-class cutlery for lifting out joints and carving out portions. Since the doorway was the only source of light and the fire the sole means of cooking, the downhearth will have been kept burning through the day, no matter what the season.

Roundhouses had no chimneys, and this influenced domestic life. Smoke rose and seeped through the conical thatched or turf roof but windless days with no draught could make for an eye-watering interior. An upper layer of smoke will have gathered under the roof, and in modern reconstructions, the smoke-free zone was found to be well below standing height. This meant that seating had to be kept low, benches and stools only a few inches high as they simply lifted sitters off the floor and not much more. Sparks from the fire needed to be watched as they spiralled up to the thatched roof, but again modern reconstruction has shown that the constant fire created a cone of carbon monoxide above head-height which usually snuffed them out before they reached anything combustible.

Around the outside of the houses ran a circular drainage ditch designed to catch the rain runoff from the roof. Along with a paved entrance at the doorway, this arrangement attempted to keep the mud manageable in wintertime. Archaeologists have found that where a cluster of roundhouses had been built, a connected series of surface drains had been dug. This was especially important if beasts were brought in in bad weather or for regular milking. In the late Bronze Age a variant known as ring-ditch houses began to appear. Instead of a central hearth, there was a mounded area in the middle and an internal circular ditch that was often paved. These were almost certainly byres used to overwinter cattle in an arrangement of radial stalls where they could be tethered overnight to munch whatever forage was available. The paved area kept their hooves out of mud (bovine mud fever can be fatal) and made it easier to collect their valuable muck. As cereal cultivation expanded after *c* 1500 BC and inbye fields were enclosed to be used again and

again, farmers increasingly appreciated the value of fertil[...]
as well as animal. Some of the larger ring-ditch house[...]
have had an upper floor where families could live. In th[...]
a bitter winter as the winds whistled round the thatch, wa[...]
the beasts will have convected upwards and the snuffling, [...]
company of the cattle may have been comforting. On the other side
of that relationship, the closeness of people and animals also meant
diseases like tuberculosis were more easily able to jump the species
barrier.

Loving Cattle

The word 'cattle' was not originally used to describe bovine
animals. It is cognate to chattel, or personal property and it
derived from the Old French 'catel' which in turn came from
the Latin 'caput'. It meant moveable property and since cattle
often formed the most valuable element of anyone's property,
a process of transference took place. In Celtic societies, cattle
were a measure of wealth and in Scots Gaelic the fondest term
of endearment is 'm'eudail', literally 'my cattle', or perhaps 'my
chattels'.

Roundhouses were probably snug enough, good places to sleep
and store supplies, but they were too dark for any close work like
weaving or even food preparation. When the weather allowed,
people lived their lives outside. Until the Industrial Revolution,
most of Britain's housing was little better lit. For example, croft-
ers' houses in the Highlands were arguably less sophisticated than
roundhouses and their interiors were so dark that they were called
blackhouses. The image of an old lady sitting at her spinning wheel
outside a Highland croft house is an ancient and valid one.

The survival of objects in the archaeological record is often a
matter of accident, but in the case of the late Bronze Age a surprising cultural habit has meant that some finds were more predictable.
Towards the end of the second millennium BC, powerful people or
their proxies began to throw away large quantities of their most
precious possessions. All over Britain, particularly in watery places

ɔuch as lakes, bogs and rivers, bronze swords, spears, shields and other war gear has been found, apparently deliberately deposited. In a religious rite of some sort, maybe out of a belief that the gods resided in lakes or rivers, these prized items were consigned to the waters – and not retrieved. So much prehistoric metal has been reclaimed from the bed of the Thames that some historians believe that it must have been seen as a sacred river. Others think that these material sacrifices were acts of propitiation, attempts to cultivate the favour of the gods, pacify their malign instincts and prevent calamity. Perhaps they are right. And perhaps the extraordinary habit of throwing beautifully worked weapons into water is echoed in the tale of Excalibur and the Lady of the Lake who caught it, her arm 'clothed in white samite'.

Some time in the late twelfth century BC an Icelandic volcano, Mount Hekla, erupted, sending 7.3 cubic kilometres of ash rocketing into the atmosphere. An extremely severe episode, scoring 5 on the Volcanic Explosivity Index, its profound impact on the climate was recorded in the Greenland ice cores. Another measure corroborates this as tree rings in Irish bog oak narrowed dramatically for 18 consecutive years. Using that and other evidence, some scientists have dated Hekla very precisely to 1159 BC. As the volcanic dust screened the sun and temperatures plummeted, agriculture suffered.

It is very likely that the roar of the eruption was heard in northwest Britain and volcanic ash deposits have been detected at several archaeological sites. In 1693 Hekla rumbled once more, this time with a VEI score of 4, and ash was blown eastwards to fall over much of Norway. In that year the volcano continued to erupt for seven months. If the 1159 BC episode, more severe at a VEI of 5, had continued as long, the effect was probably to block out the sun for several years. Famine will have followed and at the end of the second millennium, the population of Britain appears to have declined. Nearest to the eruption, the north-west will have felt the brunt. Palaeobotanists working in Caithness detected a steep decline in pine tree pollen in anaerobic cores dated to the time of the catastrophe. It dropped from 20% to 2%. Colder and wetter weather in Shetland, Orkney and the Hebrides encouraged the spread of peat and in some places the new field boundaries set up by Bronze Age farmers were submerged across the span of only a few generations. Upland cultivation gave way to stock-rearing and the population dwindled. Some historians believe that the eruption of Hekla triggered a migration from north-western Scotland to the east but at present there exists no firm DNA evidence to support this. The differences detected in the east/west frequencies of the sub-groups of R1b appear already to have been establishing themselves.

Where DNA does underline the sustainability of stock rearing as a staple for prehistoric communities is in the adaptation to milk drinking. Most mammals, including human mammals, lose the ability to digest milk once they have been weaned. But, about 10,000 years ago, when farming first arose in the Near East, a genetic variant arose that allowed older children and adults to continue to metabolise lactose, the sugar in milk. A small number of independent variants arose in pastoralist communities in Africa, but elsewhere across these vast regions animal milk is not drunk in any quantity.

In the early farming communities of the Fertile Crescent, the first people in whom this new genetic variant appeared had a great advantage, in that their bodies were able to use the milk of domesticated goats, sheep and cattle as a reliable and regular source of food. A process of natural selection began which meant that the First Milk Drinker's advantage usually ensured that he or she had many children who inherited the variant and survived in turn, they themselves having many children. In this way the ability to drink milk spread very rapidly. Anyone who carries the marker of what geneticists call lactase persistence is a descendant of the original Near Eastern farmer in whom it arose.

The modern frequency of this variant builds from east to west across Europe to a peak in the north of Britain – and that peak is clearly a reflection of the overwhelmingly pastoral economy of the Highlands and Islands over the last 3,000 years or so, and it may possibly be a legacy of the devastating effects of Hekla. Those who had inherited lactase persistence had a much better chance of survival, and natural selection has driven up the proportion of British people who have it to over 90%. In Europe it can be much lower and, for example, only 40% of Croatians can digest milk. And famously, Ötzi the Iceman was lactose intolerant.

After Hekla roared and its effects wore off, the climate clearly recovered. There is some evidence that it actually improved and that the population increased. Pressure to feed more mouths and to bring more land into cultivation drove farmers to plant crops in upland areas and the gradual realisation that it was possible for barley and wheat to ripen at altitude suggests longer summer growing seasons and higher average temperatures. Estimates of prehistoric populations are notoriously difficult to arrive at, but by the end of the first millennium BC there may have been as many as three to four million people living in Britain. As now, most were in the fertile lowlands of the south, with the least in the wilder highlands of the north.

As part of a natural human impulse to impose order, to tidy up the inconsistencies and contradictions of the past and organise history into neatly labelled phases or ages, historians have traditionally written of our prehistory in relation to weapons and tools. The Stone Age was followed, briefly, by the Copper Age and then the Bronze Age. Based on a chronology largely determined by dated archaeological finds, the Iron Age came next. But while these changes did matter and did affect the lives of ordinary people, they will not have been aware of cultural transitions as they happened. Aside from calamities like Hekla and their immediate effect on climate, change was usually imperceptible. In high contrast to our information- and communication-saturated lives, where the world around us seems constantly in flux, the brief spans of most of our ancestors will have seemed little different from one generation to the next.

Cultural change was much more evident amongst the elite, the retinues, warbands and families of men who might be called kings. To the north of the Alps, probably from the same region of central Europe as the Amesbury Archer, a new form of metalworking was developing. The skills of smelting iron ore were being perfected alongside bronze-smithing, and from the eighth century BC onwards what is known as the Hallstatt culture (named after the site near Salzburg in Austria) began to produce weapons forged in iron. Much more readily available than copper and rare tin, iron ore could be smelted and worked by itself, without the need to add another metal to make an alloy.

Like bronze, iron was hammered on an anvil into hardness.

Copper could only be made hard by the admixture of tin, but sword blades of pure iron were beaten into shape and sharpness. As techniques developed, smiths realised that the addition of carbon in the form of charcoal during the smelting process would create a very hard alloy which was an early form of steel. Blades could then be honed to a ferocious razor-sharpness and they were much less likely to shatter in battle or be bent out of shape.

In the archaeological record, the most striking effect in the transition to ironworking was the scale of the output of forges. By the middle of the first millennium BC, many swords and daggers made by the Hallstatt smiths had found their way to Britain. And after trade with Europe appeared to shrink in the fifth and fourth centuries BC, native metalworkers began to produce iron artefacts from the deposits of ore that could be mined in many parts of Britain. The strong impression is of a well-armed elite in control of a society under pressure to produce food through intensive farming.

Archaeology from the first millennium BC is eloquent about the styles and forms of landholding. Britain appears to divide into three zones. In the Atlantic west, from Cornwall to the Northern Isles, the remains of many small, defended homesteads have been found, particularly in areas where the land was cultivable and therefore more valuable. Usually each would have supported a single family of farmers and herdsmen. Perhaps the most spectacular example of the Atlantic fortress farmsteads was the broch.

Sometimes known as an Atlantic Roundhouse, these windowless stone towers were found in great numbers in the north and west of Britain. 'Broch' is from the Scots word 'brough' for a settlement or town, and although their bluff and imposing high walls strongly suggest the stamp of a fortress, they were probably only a stone version of a roundhouse that developed in a landscape where timber was very scarce.

The sites and remains of 571 brochs have been found in Scotland and most appear to have been built over a relatively brief period, between the beginning of the first century BC and the end of the first century AD. Some archaeologists believe that the similarities of structure and technique as well as the short time-frame suggest that brochs were built by specialists, itinerant teams of masons.

Perhaps the most sophisticated examples of drystane architecture ever made (where no mortar or other binding agent is used), brochs resemble miniature versions of the cooling towers around power stations and nuclear plants. Constructed with a double skin of masonry tied together with cross-slabs to make them rigid, strong as well as draught- and rain-proof, they vary a great deal in size. The smallest measure 5 metres in diameter and the largest 15 metres. Mousa in Shetland (where 78 brochs have been certainly identified and 42 suspected) is the best preserved with its walls rising to 13 metres in height.

Interpretations vary, but the consensus is that these remarkable structures were roofed by a timber-framed cone covered by thatch. On the inside walls scarcement stones protruded to support the beams of a first floor. All have a single entrance, often low (probably to force any intruder to crouch or a visitor to assume an attitude of reverence) and usually framed by a massive lintel. At Dun Telve in Glenelg, on the mainland opposite Skye, the lintel would have taken a huge effort to raise up and set in the correct position. Inside the doorways of many brochs, there is a small cell between the outer and inner walls whose function is not well understood.

Brochs are very beautiful, and that may have been the point. Made from local stone, rising out of the landscape, brought into being by human hand, they have a distinctly sculptural quality. Many stand on dominant sites that can be seen from a distance. Mousa is on a headland and on the Isle of Lewis, Dun Carloway was built on a prominent ridge overlooking a valley. Apparently impractical as forts, with little or no resource for attacking an assailant, brochs were more likely useful as temporary refuges. Perhaps they functioned like the bastle houses of north Northumberland, where sixteenth-century farmers could keep their cows safe from raiders on the ground floor while they hoped to outlast an attack as they cowered on the first floor. These sturdy buildings often had no windows. But it may have been the beauty of brochs and the conspicuous skill, and cost, that it took to build the first high-rise accommodation in Britain that mattered. Perhaps they were, as archaeologists sometimes remark, 'statements in the landscape', statements of power and prestige.

On the eastern, North Sea side of Britain, from the Thames Estuary up to the Firth of Forth, the emphasis was different and appears to have been less on defence and more on accessibility. Open villages, some of them large, are found and in the countryside beyond, there were many single farmsteads. This is perhaps surprising, given that the three to four million population of Iron Age Britain was concentrated in the south and east, where most of the best and low-lying land lay. There must have been competition, almost certainly conflict, and in many places long stretches of ditching and banks appear to mark boundaries. Archaeologists have traced the marks of a successful agrarian society: large roundhouses with hayricks on stilts nearby (presumably to keep air circulating and vermin at bay) as well as granaries and underground storage tunnels known as fogous in Cornwall and elsewhere at souterrains.

There are no written records for Britain before the voyage of Pytheas from Massalia in the 320s BC and only archaeology, and occasionally DNA, can provide the meagre threads of a narrative. Between 800 BC and 60 BC, there certainly seem to have been episodes of conflict. Battles or skirmishes were often fought at river fords (perhaps locations that were convenient, traditional or agreed beforehand) because many weapons, often slighted or damaged, possibly ritually, were deposited in the deeper water near them. It may be that these were the weapons of defeated enemies, or that there was a belief that swords and spear tips forged in the alchemy of the blacksmith's fire ultimately belonged to the gods and ought to be returned to them.

Closeness to the gods appears to have been one of the impulses behind a characteristic form of earthwork architecture in the first millennium BC. From the southern coasts of England, through Wessex to North Wales and further north to the area around the Firth of Tay, lay a third distinct zone that bisected prehistoric Britain. This was the land of the hillforts, vast communal projects laboriously dug and shaped by many millions of man-hours. Perhaps the most famous, certainly one of the largest and most monumental, is Maiden Castle in Dorset. These forts were undoubtedly directed by a powerful authority able to organise and enforce the execution of these immense earthworks, people we may call kings, or perhaps priest-kings.

At either end of the central zone lie two very different forms of hillfort, with different uses and purposes. In the Scottish Border Country, the upper Tweed Valley is dominated by the Eildon Hills. The rim of an ancient volcano, there are three peaks and on Eildon Hill North a huge hillfort was dug sometime around 1000 BC. When the westering sun is oblique, the shadow of a long ditch and bank perimeter can be clearly made out as it sweeps around the heathery crown of the hill. It is more than a mile long. Enclosing a very large area, around 50 acres, the rampart had five gateways and on the flatter plateau below and to the north of the summit, more than 300 hut platforms have been found. When they are etched by a light dusting of snow, many clusters of hut circles emerge from the grass.

Eildon Hill North commands wide vistas as it looks east to the widening Tweed Valley, south to the Cheviots and north to the Lammermuir Hills, while at its western back the Southern Uplands rise. The hillfort was indeed a grand statement in the landscape, one that was visible for many miles from all directions, and especially from the eastern valley of the great river as it winds towards the North Sea. Whoever ordained that it should be made was unquestionably powerful and resourceful. But the priest-king of the Tweed was not interested in defence or anxious about an attack on his dramatic citadel.

At a mile in length and with five gateways, the rampart circling Eildon Hill North could never have been manned, far less defended. There is also no source of water anywhere on the summit and a force of any size could not have been maintained for any length of time. What caused the king of the Tweed to command the back-breaking work of hacking out ditches with mattocks, loading the upcast into baskets and piling the earth on a rampart can only be guessed at. But his motives are likely to have been both religious and political – and perhaps economic.

By the first millennium BC native British society had become recognisably Celtic in its nature, with dialects of what might be termed Old Welsh spoken over most of the island (Ireland understood itself in early Gaelic, different by that time but a cousin language). The turning points of the stock-rearing year, ancient festivals still remembered in Gaelic and Welsh, may well have been celebrated on Eildon Hill North and the hillforts of southern Britain.

Samhuinn translates translates approximately as 'the end of summer' and its modern, and more than faintly pagan descendant, is Halloween. It falls at the end of October and is usually seen as the onset of the winter. Bonfires were lit in the gathering gloom and feasting took place as beasts were slaughtered and the late fruits of autumn enjoyed. And fires may have flickered on the windy summit of Eildon Hill North at Samhuinn. About 20 miles to the west, on the banks of the young River Clyde, stands another singular hill where a fort was dug. It is Tinto Hill, and the name derives from a Celtic root that means Fire-Hill.

The ancient emphasis on stock-rearing is perhaps best seen in the most obscure and half-forgotten of the Celtic quarter days. Imbolc has been Christianised as St Bride's Day and it falls at the beginning of February. Ewes that had been tupped in the summer began to lactate in anticipation of lambing and their sweet milk was welcome nourishment for herdsmen and their families in the hungry months of the late winter. Beltane was celebrated on 1 May and it marked another turning point in the stock-rearing year. This was the beginning of the summer, time to gather herds and flocks for the drive up-country, up the hill trails to the high pastures where the new, sweet grass was peeping through the yellow wrack of winter.

Milk, butter, cheese and their central role in sustaining farming communities were all celebrated in country districts until modern times. The naturalist and writer Thomas Pennant travelled the length and breadth of Scotland, often on foot, just before the Agricultural and Industrial Revolutions swept aside the ancient rites of life on the land forever. As Pennant recorded, the old journey of transhumance could not begin without ceremony and in country districts these rituals were still unmistakably pagan in the 1770s as the people understood the rich gifts of pastoralism:

The herds [men] of every village hold their Beltane. They cut a square trench in the ground, leaving turf in the middle. On that they make a fire of wood, on which they dress a large caudle [pot] of eggs, oatmeal, butter and milk, and besides these they bring plenty of beer and whisky. Each of the company must contribute something to the feast. The rites begin by pouring a little of the caudle on to the ground by way of libation. Everyone

131

then takes a cake of the oatmeal, on which are raised nine square knobs, each dedicated to some particular being who is supposed to preserve their lands, or to some animal, the destroyer of them. Each person then turns to face the fire, and breaks off a knob, and flinging it over his shoulder, says, 'This I give to thee, O Fox, spare my lambs.'

The last of the four Celtic quarter days is still celebrated in rural Ireland. Lughnasa falls on 1 August and is often described as a harvest festival even though most modern corn growing in Britain and Ireland needs a further six weeks to ripen. But for a society based on the husbandry of sheep, goats, pigs and cattle, it may have had a different significance. The spring lambs, kids, piglets and calves will have been well fattened by August and it may be that Lughnasa was a time of tithe-gathering at Eildon Hill North. The royal authority that caused the hillfort to be built will have been sustained by the produce of herdsmen and at the foot of the three hills lies a complex of earthworks that look very much like stock enclosures. A series of ditches and banks that follow the fall of the ground, terminating at streams or other natural features, they may have penned animals due to the king of the Tweed as his royal portion.

If the great hillfort was not a fort, it may be better understood as both a seat of great temporal power and perhaps as a sky-temple. Not that the people of the first millennium BC will have made that distinction. Those who worked the land depended on the weather and their direction of worship and propitiation may well have been skywards, to where the gods swirled amongst the elements and rained down storms or radiated sunshine. It is impossible to do more than conjecture, but altitude clearly mattered very much. There can have been few more inconvenient places for government or worship than Eildon Hill. And it was not unique in the north. There exist several great hillforts dating to the early first millennium BC that can only be reached after a long and arduous climb.

Eildon Hill North often catches a scarf of low cloud even when sunshine bathes the Tweed Valley below. If those who lit fires to mark the great festivals of the farming year on the summit so that the sky-gods could see them blaze, then days when cloud descended may have hidden even more mysteries. As it wrapped itself around the heathery flanks of the hill, shifting, drifting and fading, it may have been seen as a sacred mist.

In the south it was probably different. For many seasons through the 1970s, the hillfort of Danebury was excavated by teams led by one of Britain's very greatest prehistorians, Barry Cunliffe. His people meticulously dug more than half of the area of the 12-acre site and early on they established that the single ditch and bank that enclosed it were made some time around 600 BC. Over a life lasting half a millennium, the fort acquired another ring of ditches and banks, and these may have protected a grassy corridor where livestock could have been securely held in the event of an attack.

Built on a prominent hill looking out over the undulating farmland of Hampshire, about 12 miles north-west of Winchester, Danebury was the stronghold of a powerful individual, perhaps a king, certainly a local magnate. Other hillforts were dug in the same region at about the same time and each appears to be the focus of a clear domain, a scatter of farmsteads that probably supplied the retinues behind the ramparts. At Danebury a series of roundhouses nestled against the inside of the encircling bank for better shelter and either side of two parallel tracks were rows of four-post structures that have been interpreted as granaries. They may well have held the tithes paid by the surrounding farmers. At its height, around 400 BC, Danebury had a population of between 200 and 350. In the centre of the oval-shaped fort, structures were found that appear to have been shrines. And traces of metalworking and other manufacture on some scale were identified. These finds point to the hillfort as the power-base of a leader, perhaps someone who also had priestly status and officiated at rituals around the central shrines.

Unlike Eildon Hill North, Danebury really was a fort. In places, the height of the rampart was awe-inspiring at 16 metres from the bottom of the ditch to the top of the palisade. In contrast to Eildon Hill where there were five, there were only two well-protected gateways and they were accessible by a serpentine approach that would have exposed attackers to a barrage of missiles before they even reached the timber defences. Barry Cunliffe's excavators found clear evidence of conflict. Both of the gateways of Danebury had been burned down and towards the end of the fort's life, some time around 100 BC, the east gate had again been attacked and set alight.

Charnel pits were also uncovered and these contained about 100 bodies that had all been buried at the same time, almost certainly in the wake of an attack on Danebury. The skeletons of many were found to bear the marks of ancient sword and spear wounds. There is a sense of the Hampshire hillfort as a prototype of a much later medieval castle, a well defended place of refuge for the surrounding countryside, its farmers and their animals. By the end of the first millennium BC, Danebury appears to have dwindled dramatically in importance and its ramparts dwarfed what was probably only a single farmstead.

In 55 BC the farmers of Hampshire will have heard startling and worrying news from the east. For months rumours had been circulating, but in late August that year, they turned into a political reality. Warned by scouts that a Roman invasion fleet had set sail from Boulogne, British kings mustered their warbands and rode hard for the coast to meet it and its far-famed commander, Gaius Julius Caesar. From the heights of the white cliffs of Dover, the kings of the Catuvellauni and the Cantiaci could make out the sails of about 80 troop transports. Their scouts will have made hurried estimates – maybe 7,000 or 8,000 men, maybe fewer. Across the waves of the Channel, battlehorns blaring, swords and spears clashing against their shields, the British roared defiance.

Spurring on their horses and flicking a whip across the backs of their chariot ponies, the British army moved quickly eastwards, guessing that the Romans would seek to drive their ships up onto the shingle beaches near Deal. The invasion cannot have been a surprise. There is some evidence that Caesar was considering launching it in 57 BC or 56 BC. Commius, King of the Atrebates, a kindred whose lands lay south of the Thames (Danebury was on its western borders), had been despatched as an ambassador to encourage negotiations with Rome. Instead, the Atrebatean had been arrested and no doubt details of his allies' invasion plans were extracted.

On the afternoon of 27 August 55 BC, watched by the British army, Caesar ordered his captains to drop anchor in the roadsteads off Dover. A council of war was summoned aboard his flagship. The weather in the Channel had been changeable and before any further advance could be contemplated, the whole Roman fleet needed to come together so that any landing could have maximum impact.

Somewhere north-east of Deal, Caesar signalled that his fleet should turn and steer shorewards to run their keels aground on the shingle beaches. The British kings and their charioteers had probably outrun their infantry in their dash along the coast, but they could still mass in numbers to repel to Roman landing. This was the moment of greatest danger for Caesar – and his cohorts were almost driven back into the sea. While the war galleys with their artillery, crossbows and slingshots had been rowed hard enough for the shore so that they could beach, the heavier transports could not get close enough to disembark the legionaries. Their draught was too deep, and with 100 men on board, they were too heavy to be driven up onto the shingle. Caesar left a record of what happened next:

> And then, when our soldiers were still hanging back, mainly because of the depth of the water, the standard-bearer of the Tenth Legion offered up a quick prayer and then shouted out, 'Jump down, soldiers, unless you want to give up your eagle to the enemy; everyone will know that I at least did my duty to the Republic and my commander!' After saying this in a loud voice he jumped off the ship and began carrying the eagle standard towards the enemy. Then our soldiers called out to each other not to allow so terrible a disgrace [as to lose the standard] and leapt down from the transports. When those on the nearby ships saw them, they followed and began to close with the enemy.

Directing operations from his flagship, Caesar could see where the shield-wall of the legionaries' battle-line was too sparse or about to buckle as the British warbands attacked the beachhead. In rowing boats, parties of soldiers were sent to reinforce weak points, and as Roman discipline locked the line together, their determination drove them up the shore to gain a foothold. As the battle wore on and more troops were safely landed, the British kings signalled a retreat. Caesar's famous luck had held, for the moment.

The cavalry transports had not boarded at Boulogne on 26

August and even after the victory at the beachhead, they still had not made landfall on the Kentish coast. That meant no pursuit of the fleeing British army was possible and no opportunity to inflict the heavy casualties always sustained by men in flight. Caesar was therefore forced to secure only the immediate hinterland and wait for his delayed cavalry squadrons.

Diplomacy did finally pay dividends. The kings of the Catuvellauni and Cantiaci sent envoys and freed Commius, probably in the hope that he could mediate. Once his cavalry landed, Caesar knew that his bargaining hand would be strengthened. But for days after the beachhead had been gained, lookouts saw the cavalry transports blown offshore and back to Gaul by a storm. It also refloated the troop transports so that they broke their moorings and smashed into each other or were swept out to sea. Suddenly Caesar's luck looked to be wearing thin and prospects were transformed.

Seeing the Romans stranded in their camp on the coast and suddenly exposed, the British kings broke off negotiations, mustered their hosts and advanced. After a bitter, close-run but victorious battle at the beach-camp, Caesar bought enough time to have the transports repaired. Cramming the legionaries on board, he led his expedition back across the Channel to the safety of Gaul in what looked more like a retreat than a return in triumph.

But events were spun very differently in Rome. Great Gaius Julius Caesar had led his victorious legions to the very ends of the Earth, crossed the dangerous waters of the Ocean (little did the crowds in the Forum know) and defeated the barbarian hordes. An unprecedented 20 days of public thanksgiving were declared and Caesar's enemies in the Senate were silenced. The reality of a series of narrow escapes was successfully suppressed.

The expedition of 55 BC had left many loose ends, but it also offered an opportunity. A year later Caesar was back in Britain and this time he may well have intended a wholesale conquest of at least part of the island and the creation of a new province, just as he had done in Gaul. It would bring enormous riches and the accompanying prestige, the glow and glory of conquest, would consolidate Caesar's supremacy amongst the people and quieten the whispering, plotting porticos of power in Rome.

Lessons were learned. Transports that could be beached more

easily were built. They had a shallower draught and banks of oars to add propulsion. Six hundred were commissioned to carry a much larger expedition, an invasion force of four legions and 1700 cavalry troopers. This time real territorial gains were made. After a battle probably fought around the hillfort at Bigbury on the River Stour near Canterbury, Caesar led his army to the banks of the Thames and crossed. With his supply-lines extended and his ship-camp of transports far to the south-east, it was time for diplomacy.

Cassivellaunus, king of the Catuvellauni, had harried and weakened the Roman army as it had advanced through his realm. He had removed the neighbouring king of the Trinovantes, a kindred based around the modern county of Essex, and Caesar attempted to play on this local dispute. Mandubracius, heir to the deposed king, had accompanied the Roman expedition and in what seemed like a premeditated move, he persuaded his countrymen to ally themselves with Caesar. As a guarantee, they gave up hostages and agreed to supply the legions with much-needed food. They also betrayed the location of Cassivellaunus' headquarters and when the Romans broke through its defences, they found not the king but a prize almost as welcome, herds of his cattle.

Despite an attack led by the four kings of the Cantiaci on the Roman ship-camp, the British confederacy decided to sue for peace. And Caesar must have been glad to negotiate. Time was pressing, it was late September and messengers had brought word that rebellion was brewing across the Channel in Gaul. Outline terms that included the usual condition of hostages, an amount of tribute and a guarantee of security for the Trinovantes were quickly agreed and the Romans embarked once again for Gaul. This time reaction in Rome was more muted. Quintus, the brother of the great orator and politician, Marcus Tullius Cicero, had been an officer in Caesar's army and he wrote letters home that were brimming with disappointment at the lack of booty 'apart from captives, and I fancy you won't be expecting them to be highly qualified in literature and music'. It would be another 90 years before legionaries would set foot once more on British soil.

Unusually for the time, the Romans were a literate culture whose historiography was intimately connected to politics. Caesar wrote of his victorious wars in Gaul and Britain both to glorify his achievements and also gain a measure of control over the narrative, something politicians still strive to do. But his work also supplies, in many cases, the first written record of the peoples he subjugated.

What is immediately striking in the accounts of the expeditions to Britain are the links between what is now northern France and Belgium with the British kingdoms of the south-east of England. Famously, Julius Caesar wrote that Gaul was divided into three parts: Aquitania in the south-west, Celtica or Gallia in the centre stretching from the Atlantic coast to the Alps, and Belgica in the north. Archaeological, historical and genetic evidence points to continued contact between Gaul and southern Britain, but especially with Belgica. In the first century BC this region included most of north-eastern France, bordering the Seine in the south and the Rhine Delta in the north. Deriving from an Indo-European root cognate to 'bulge', Belgae has the sense of swelling with rage, or battle-frenzy. The Romans certainly respected Belgic warriors.

Caesar described the kindreds of the north-east of Gaul as both Celtic and Germanic. From the evidence of personal names and place names, their language appears to have been Celtic, however their links with the Rhineland and beyond appear to have been close. Here is Caesar again:

> The greater part of the Belgae were sprung from the Germanic peoples, and that, having crossed the Rhine at an early period, they had settled there [in Belgica] on account of the fertility of the country.

There is clear evidence that they also migrated in numbers across the Channel to Britain. From *c* 175 BC Gallo-Belgic gold coins began to be minted for use not as currency but as gifts, and these found their way quickly to the south-east, probably via trade up the Thames. It seems that an entire kindred invaded, perhaps during the first century BC, and occupied the territory around Winchester, extending their control from the Severn to the Channel coast around Southampton. Caesar commented that the Belgae 'came to raid but stayed to sow'.

When the Romans later attached placenames to the map of their province of Britannia, what would become Winchester was labelled

Venta Belgarum, the Market-Town of the Belgae. Historians also believe that other kindreds were taken over and governed by a Belgic warrior aristocracy and that the federation of the Catuvellauni, the Cantiaci and others had cultural as well as political and military links. Diviciacus, King of the Suessiones (who gave their name to Soissons) also had dominions in Britain, and Commius, who changed sides at least once, fled from Caesar's conquest of Gaul 'to return to his own people in Britain'. And kindreds with the name of the Atrebates lived on both sides of the Channel.

Archaeologically and historically the links are clear, and DNA also shows that people with Germanic markers had crossed the southern North Sea and the English Channel from at least the first century BC onwards. It was the beginning of a long period of genetic traffic.

6

On the Edge of Beyond

✖

NO ONE HAD EVER seen anything like them. Their grey shapes looming out of the land of nightmares, monsters twice the height of a man plodded into the centre of the settlement. On their backs sat men endowed with magical powers. Using only a stick, tapping the flanks of the monsters, they could make them stop or walk at will. As the native peoples gawped and even their defeated warriors shrank back, the Romans brought elephants into Britain.

It was the late summer of AD 43 when the Emperor Claudius came to Camulodunum, modern Colchester, at the head of his victorious legions. As well as war-elephants, there were impressive pieces of artillery such as catapults and cross-bow bolt launchers, all designed to overawe. Before the age of mass communication, show mattered and Roman emperors understood the central importance of making the instruments of their power visible to as many as possible in military parades, straightforward shows of strength.

The arrival of the elephants and the emperor were the crowning moment of what had been a conquest of south-eastern Britain undertaken for essentially political reasons. Claudius badly needed the kudos of military success and memories were still fresh of how the Roman mob had marvelled at the audacity of Julius Caesar's expedition to the edge of the world. What better way of confirming Claudius' credentials as a strong leader than to have him ride at the head of a conquering army that had dared to cross the Ocean and subdue barbarians? But the whole enterprise had very nearly gone badly wrong before a single troop transport set sail.

Writing more than 100 years after the invasion, Lucius Cassius Dio compiled *Romaika*, a history of Rome in 80 books. Originally in Greek, it is especially authoritative on the early years of the Roman province of Britannia. Here is Dio's account of the somewhat surprising prelude to the expedition as the legions mustered at Boulogne. Faced with a mutiny, the commanding general, Aulus Plautius, was forced to send for Narcissus, a freed slave who was the imperial chief of staff:

Thus it came about that Plautius undertook this campaign; but he had difficulty in inducing his army to advance beyond Gaul. For the soldiers were indignant at the thought of carrying on a campaign outside the limits of the known world, and would not yield him obedience until Narcissus, who had been sent out by Claudius, mounted the tribunal of Plautius and attempted to address them. Then they became much angrier at this and would not allow Narcissus to say a word, but suddenly shouted with one accord the well-known cry, 'Io Saturnalia' (for at the festival of Saturn, slaves don their masters' dress), and at once they willingly followed Plautius.

Roman soldiers disliked the thought of sailing the Ocean, the name they gave to the English Channel, but it seems that the joke made at Narcissus' expense broke the tension and they at last allowed their centurions to persuade them to board the transports. Four legions, about 20,000 men, crossed with as many auxiliaries, an invasion force of more than 40,000 battle-hard troops. They met a British army commanded by two princes, Caratacus and Togidubnus (sometimes known as Cogidubnus), the sons of the late king of the Catuvellauni, Cunobelinus, Shakespeare's Cymbeline. Probably on the banks of the River Medway, near Rochester in Kent, the decisive action in the war was fought. For two days battle raged until it was turned by the bravery of the tribune, Hosidius Geta, and his men. The British retreated to the line of the Thames and, probably as part of a pre-arranged plan, Aulus Plautius sent messengers back across the Channel to summon Claudius to Britain. Dio takes up the sequence of events:

When the message reached him, Claudius . . . set out for the front . . . he came to the Ocean and crossed over to Britain, where he joined the legions that were waiting for him near the Thames. Taking over the command of these, he crossed the stream, and engaging the barbarians, who had gathered at his approach, he defeated them and captured Camulodunum, the capital of Cunobelinus. Thereupon he won over numerous tribes, in some cases by capitulation, in others by force, and was saluted as imperator several times, contrary to precedent; for no man may receive this title more than once for one and the same war. He deprived the conquered of their arms and handed them over to Plautius, bidding him also to subjugate the remaining districts. Claudius himself now hastened back to Rome, sending ahead the news of his victory by his sons-in-law, Magnus and Silanus. These, on learning of his achievement, gave him the title of Britannicus and granted him permission to celebrate a triumph. They voted also that there should be an annual festival to commemorate the event and that two triumphal arches should be erected, one in the city and the other in Gaul, because it was from that country that he had set sail when he crossed over to Britain.

Behind all of these seemingly co-ordinated events, the hidden hand of diplomacy is more than hinted at. According to Dio, the pretext for invasion was the reinstatement of Verica, an exiled king of the Atrebates. But many other contacts will have been made down the length of Britain. An inscription retrieved from Claudius' triumphal arch in Rome records that at Colchester, the emperor received the submission of no fewer than 11 British kings. And one of them had come a very long way.

When the fourth-century historian Eutropius made a list of all of the emperors of Rome and their achievements, he noted next to Claudius' name that 'he added to the empire some islands lying in the Ocean beyond Britain, which are called the Orkneys'. Long thought to have been an error or an exaggeration, archaeology recently confirmed this remarkable assertion. At the impressive broch-village of Gurness on Orkney's Mainland, sherds from a Roman amphora were found. It was a type that had become obsolete by AD 60 and it had contained a fancy, sweet liqueur enjoyed by aristocrats. It seems likely that the Orkney king was indeed at Colchester, had seen the elephants and bowed down before Claudius to pledge the submission of his kingdom on the world's edge. And perhaps he was given a gift of wine from Italy to take home.

The emperor was in Britain for only 16 days before he hurried south to Rome to make the most political capital possible out of his triumph over the barbarians. With such a tight timetable, the arrival of all 11 kings in Colchester must have been organised in advance and stage-managed for maximum effect. And not only did the coming of the Orcadian king speak of effective and rapid long-range communication, it also allowed Claudius to claim unparalleled reach – to the very ends of the Earth. It was a propaganda master-class, probably orchestrated by Narcissus, the freedman who suffered the contempt and scorn of the mutinous legionaries on the beach at Boulogne.

After the elephants had been led onto transports to cross the Channel, Aulus Plautius and his four legions began to push the boundaries of the new province. A future emperor, Vespasian, marched westwards with the II Augusta and attacked 20 hillforts, including Maiden Castle, where a series of pits or war-graves have been found. Fighting must have been fierce. By the late AD 40s all of the fertile south-east had fallen under Roman control. The frontier was marked by the Fosse Way, a road rather than a barrier because above all Roman generals prized mobility.

All roads not only led to Rome, the beating political and economic heart of the empire, they also drew Britain more closely into a network of European communication and trade. As the legions extended the frontiers of the new province of Britannia to Wales and southern Scotland, much of the native population became exposed to a wider world. The empire offered Britons opportunities and Britain became a destination for outsiders.

In the midst of the political and military ferment that followed the invasion of AD 43, it might reasonably be expected that the sum of Britain's DNA saw substantial augmentation and alteration. Estimates of the population of the province vary widely, from 1.5m to 3.6m at the peak of its prosperity in the third and fourth centuries. When Claudius and his elephants plodded into Colchester, it seems reasonable to believe that around 2 million people lived in what is now England and Wales. It is against this context that the arrival of the legions and the regiments of auxiliaries should be seen.

In the decades immediately after the conquest 40,000 soldiers and cavalrymen, a huge proportion, perhaps an eighth of the entire imperial army, were stationed in Britain. Because they were needed. In AD 60 the new province erupted in rebellion.

It was Roman policy to govern by proxy where possible and in Britannia, the kingdom of the Iceni was ruled by a client, Prasutagus. When he died, his wife and heir, Boudicca, offered to continue the arrangement, formally sharing her territory, what is now Norfolk and part of eastern Lincolnshire, with the Emperor Nero. But at that moment Roman attitudes may have intervened. In Celtic Britain queens governed substantial realms. The largest kindred of all, the Brigantes of the north, was ruled by Cartimandua in her own right. But it seems that the Roman provincial government simply could not countenance the idea of Boudicca sharing her throne, or any throne, with Nero.

When the queen of the Iceni objected, reaction was brutal. She was stripped and whipped with rods, perhaps publicly, while her daughters were raped by soldiers. Wealthy Icenian aristocrats were expelled from their estates and cash payments thought to have been bounties from the emperor turned out to be loans. And these were recalled with immediate effect.

After her humiliation and the vicious defilement of the princesses, Boudicca wasted no time in raising rebellion. Here is Dio's amazed description of the British queen, a mere woman:

> But the person who was chiefly instrumental in rousing the natives and persuading them to fight the Romans, the person who was thought worthy to be their leader and who directed the conduct of the entire war, was Boudicca, a British woman of the royal family and possessed of greater intelligence than often belongs to women. This woman assembled her army, to the number of some 120,000, and then ascended a tribunal which had been constructed of earth in the Roman fashion. In stature she was very tall, in appearance most terrifying, in the glance of her eye most fierce, and her voice was harsh: a great mass of the tawniest hair fell to her hips; around her neck was a large golden

144

James D. Watson on the left and Francis Crick with their model of the DNA molecule. (E. Barrington Brown/Science Photo Library)

Maurice Wilkins. (Science Photo Library)

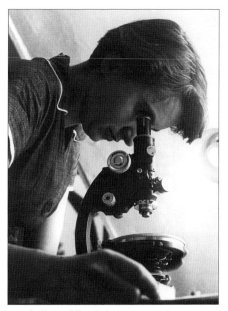

Rosalind Franklin. (Science Source/ Science Photo Library)

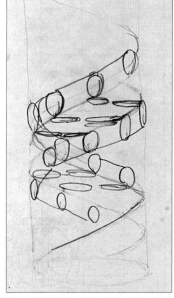

Francis Crick's original sketch of the double helix. (Science Source/ Science Photo Library)

Above. The Great Rift Valley of Tanzania. (FransLanting/ Mint Images/ Science Photo Library)

Right. The Olduvai Gorge in Tanzania. (Javier Trueba/ MSF/Science Photo Library)

Louis Leakey measures the fossil-bearing strata in the Olduvai Gorge. (Des Bartlett/Science Photo Library)

Richard Leakey. (Science Source/ Science Photo Library)

Excavations in the Denisova Cave.
(Ria Novosti/Science Photo Library)

What Neanderthal men looked
like. (S. Plailly/E.Daynes/Science
Photo Library)

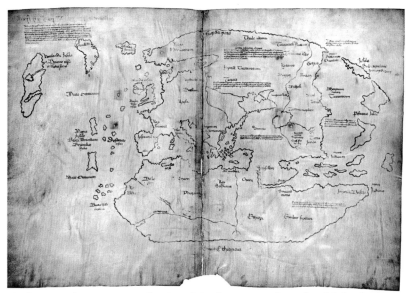

The famous 15th-century map showing Vinland. It apparently predated Columbus'
voyage of 1492.

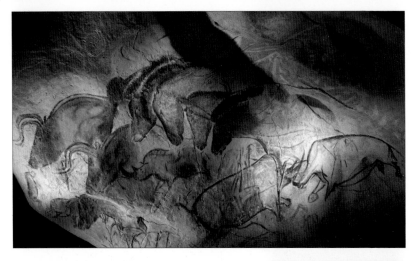

Above. The gallop of horses, cattle and rhinoceros across the walls of Chauvet Cave. (Javier Trueba/MSF/Science Photo Library)

Right. What Cheddar Man may have looked like.

In the foreground the Ring of Brodgar, beyond it the Ness of Brodgar and across the causeway the site of the Stones of Stenness and Barnhouse. (copyright © Richard Welsby. Licensor www.scran.ac.uk)

The magnificent Stones of Stenness. (copyright © Richard Welsby. Licensor www.scran.ac.uk)

A misty winter sun over Stonehenge. (English Heritage)

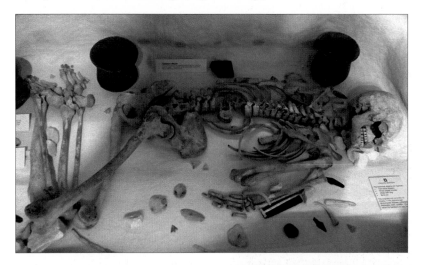

Above. The Amesbury Archer and his grave goods.

Right. Jacob Grimm. (Mary Evans Picture Library)

Eildon Hill North with a dusting of snow showing the hut circles near the summit. (copyright © RCAHMS Licensor www.scran.ac.uk)

A reconstruction of part of Hadrian's Wall. (copyright © Liz Hanson)

A slim version of William the Bastard as he stepped ashore to become the Conqueror. (Mary Evans Picture Library)

Grubbing for edible potatoes in the Irish Famine. (Mary Evans Picture Library)

Bede shown with a bee on his sleeve, a symbol of diligence and indefatigability. (Mary Evans Picture Library)

Mousa Broch, Shetland. (copyright © Historic Scotland. Licensor www.scran.ac.uk)

Maiden Castle in Dorset. (English Heritage)

Complete with tawny hair, spear and grassy tribunal, Boudicca decorously foments bloody rebellion. (Mary Evans Picture Library)

Overgrown but still impressive, Offa's Dyke. (English Heritage)

"IMMIGRANT JEWS."

Jewish immigrants arriving in early 20th-century London. (Mary Evans Picture Library)

Above. Posed in front of a Union Jack, Enoch Powell. (Getty Images)

Right. Examining a DNA sequence.

torc: and she wore a tunic of diverse colours over which a thick mantle was fastened with a brooch. This was her invariable attire. She now grasped a spear to aid her in terrifying all beholders.

No doubt swelled by the ranks of the disaffected and the opportunistic, Boudicca's host descended on Colchester. The town was set alight and its population slaughtered. Her progress southwards was not impeded because the Roman governor, Suetonius Paulinus, was in north Wales attacking the island of Anglesey with most of the imperial garrison. Rampaging on to London, the Icenian army swept all before it, and burned the city before turning north-west to meet the legions hurrying back from Wales. Stories of appalling atrocities began to circulate. Female Roman captives, or possibly women who had married Romans, were tied to trees, their breasts were cut off and stuffed into their mouths. Others were said to have suffered the dreadful agonies of impaling where a sharpened stake was forced into their anus and shoved up through their bodies.

On the Roman road known as Watling Street, perhaps somewhere near Towcester, the armies met. Once again the iron discipline of a smaller but better led Roman army triumphed. Boudicca committed suicide and over the east and south of England, smoke rose over native settlements as reprisals began.

After the carnage of AD 60, Roman Britain appeared to settle as the legions pushed northwards to establish a frontier between Carlisle and Newcastle. And to achieve further conquest and also as a deterrent to further uprisings, a heavy military presence was maintained. Consequently, it is worth considering the demographic impact of the army alone. Forty thousand soldiers and more bureaucrats must have been more than highly visible. Merchants and others came behind the army and over a matter of only a few years, perhaps as many as 70,000 new people had settled in the province, 3.5% of the total population. And almost all of them were men. That prompts another statistic, and it may be that 7% of the male population of Britain in the decades after AD 43 was made up of incomers.

Their impact was felt in at least three ways. Outside the walls of permanent forts and legionary fortresses such as York, Caerleon and Chester, civilian settlements quickly grew up. Known as 'vici', they housed shopkeepers, tradesmen and other services. Giving reasons for his disapproval, the Emperor Hadrian outlined exactly why soldiers valued vici when he complained that they contained 'drinking booths, gambling halls and prostitutes'. In the early empire Roman soldiers were not permitted to marry, but some nevertheless took native women as wives and no doubt most paid for the services of prostitutes after an evening of drinking and dice.

The growing cities of Britain were usually garrisoned. The remains of Roman London's fort lie under the church of St Giles Cripplegate and its precinct. With a population of around 35,000 at its zenith, the city was undoubtedly a melting pot as cargoes and people were unloaded at the wharves on the Thames. York, Cirencester and Dorchester were also busy and the reach of imperial commerce can be illustrated by exotic discoveries in the north. At Carlisle the tomb of a merchant from Palmyra in modern Syria has been found and in AD 306 the fort at South Shields was renamed Arbeia. It means the Fort of the Arabs and marked the arrival of the Numerus Barcariorum Tigrisensium, the Tigris Bargemen from the Middle East.

When legionaries retired, usually after long service, they were awarded pensions and often given land. Colonies were set up for veterans and in Britain these were sited at Gloucester, Colchester, Lincoln, York and elsewhere. It is likely that substantial numbers of those who retired while reaching the end of their service in Britain will have accepted grants of land in and around the colonies, especially if they had taken native women as partners and had had children by them. In addition, the provincial government will have encouraged veterans to settle to further a clearly stated policy of Romanisation in a new province. In the first century AD legionaries were either all Italians or the descendants of Italians who had settled in the first provinces of the empire. For example, the great general and governor of Britain, Agricola, came from Provence but his family were originally from Italy. It is not known how many Italian veterans decided to remain in Britain, but the colonies thrived. And it is worth noting the scale of veteran settlement elsewhere in the empire. Between 31 BC and 2 BC the Emperor Augustus granted land to 300,000 retired legionaries in Italy, Gaul and Spain.

Traces of the DNA of the legions, the grizzled old soldiers of the II Augusta, the IX Hispana, the XIV Gemina and the XX Valeria

Victrix, and others who came to postings in the province can probably be found in the male population of modern Britain, but only if careful detective work and some caution are employed. On the basis that before the middle of the second century AD recruitment was restricted to men who were Roman citizens and therefore Italians or of Italian descent, it may be reasonably assumed that a comparison between Italian Y-DNA and British Y-DNA will show up the genetic legacy of the legions. That is, those who retired and settled in Britain or those who fathered male illegitimate children before being posted elsewhere. And perhaps as an additional by-blow, the sons of Italian provincial administrators, merchants, specialist craftsmen and other civilians associated with the imperial army.

But before looking in detail at comparisons, two important caveats need to be entered. As with Britain, DNA often arrived in Italy from elsewhere, usually from the east, and what might first look like DNA inherited from the legions may well have made landfall on British shores independently, either before or after the four centuries of the Roman province. But by no means all of it did.

One simple and logical test may help to isolate the inheritance of the empire. Since Ireland was never conquered and the south of Scotland was held only for a short time in the mid second century, Roman DNA ought to be significantly more present in England and Wales. And at least one other factor needs to be borne in mind. In the third and fourth centuries AD, provincial garrisons tended not to be uprooted and posted elsewhere in the empire. For example the Roman name for York, Eboracum, was for a time interchangeable with 'Ad Legionem Sextam', the settlement of the VI Legion. And in Britain there is evidence for the accommodation of families at Housesteads Fort on Hadrian's Wall and possibly at other military bases. It is also likely that soldiering became hereditary, with sons automatically following their fathers as recruits.

To complicate matters further, there were also occasional exotic imports and perhaps the largest and best documented was the settlement of 5500 Sarmatian cavalry (and their horses) at Ribchester Fort in Lancashire. Originating in modern Romania, these large squadrons were deliberately posted as far from their homeland as possible by the Emperor Marcus Aurelius in AD 175. There is little later archaeological trace of these eastern Europeans and they may have been recalled from Britain in a bout of imperial civil war at the end of the second century.

Five major Y-DNA markers are likely to represent at least a partial sense of the genetic stamp of the legions. But the first of these almost certainly began to arrive in Britain much earlier, around 3000 BC, and perhaps even before then. Known as Alpine, R1b-S28 is found at a frequency of 13% in Italy, 6.5% in England and Wales, 4.3% in Scotland and 1.8% in Ireland. It may have been the marker of the Amesbury Archer, the man in the grave found near Stonehenge, but its frequency in Italy and England and Wales compared to the notably lower percentages in parts of Britain never or little occupied makes it a probable candidate as a legacy of the legions.

At the 2011 census, the combined population of England and Wales was 56 million. If approximately half that number are men, and discounting the 3.25 million men in England who are foreign-born, that means a total of 1.6 million carry the Alpine marker. And also setting aside those who arrived before the legions disembarked in AD 43 or after the remnants of the Roman administration left in AD 410, that may mean that around half a million men in England and Wales are descended from the men who marched behind the eagle standards.

Now, if that calculation feels highly speculative, other factors may show it actually to be conservative. It is important to note that at least four additional Y-DNA markers may have arrived with the Romans. What have been described as the Balkan group (E-V13), the Ancient Caucasians (G-S314), the Herdsman-Farmers (J-M172 – and a sub-group of this, M67, looks particularly Italian) and the Anatolian marker (R1b-M269*), when taken together, potentially add another 2.3 million Englishmen and Welshmen who could trace their fatherlines to the veterans of the II Augusta, the IX Hispana, the XIV Gemina, the XX Valeria Victrix and the other Italians who crossed to Britain in their wake. If this, in turn, appears to be far too large a number, then two-thirds of the carriers of these four markers can be subtracted because they arrived either side of the life of the province, before AD 43 or after AD 410 (for example the first wave of farmers many of whom probably carried the G group of markers). But what even rudimentary calculations show is that many British men in the direct male line may indeed be descendants of the Romans, perhaps as many as a million, perhaps even more. No account has been taken of other, rarer groups probably brought to Britain by the Romans. These include the likes of E-V22, J-M267, E-M34 and T-M184, which add up to 2% of English men.

When the men of the auxiliary regiments who marched and rode with the legions retired from active service, they were given diplomas cast in bronze. Proof of their status as Roman citizens, some of these went into considerable detail. An example dating to AD 103 found near Malpas in Cheshire was issued to a group of veterans and it listed the names of 11 auxiliary units serving in Britain at that time. Four had been raised in the Balkan peninsula and from a recent study of modern DNA markers, it seems that a significant number of men who carry E-V13 decided to settle in Britain after discharge from the Roman army.

The bronze plaque was discovered near Deva, the legionary fortress at Chester, and in Cheshire and along the coast of north Wales, the frequency of E-V13 is unusually high. This may be so because of the likely arrival of men with the same marker to mine the copper at Great Orme Head in the third millennium BC. The distribution of E-V13 also reaches east through Nottinghamshire and it is found in a swathe from Fakenham in Norfolk to Midhurst in Sussex. Across England as a whole the Balkan marker is much more common, at 1.6%, than it is in Ireland, at 0.6%, and significantly more common than in Scotland, where 1.1% of all men carry it. And, intriguingly, E-V13 is found rarely in the Midlands of England, where there are no records of units from the Balkans ever being stationed during the four centuries of the province of Britannia.

The northern European auxiliary regiments are, by contrast, scarcely worth even a speculative stab at their genetic legacy. Many of these men who fought so well for Claudius and Aulus Plautius at the River Medway were recruited from northern European kindreds. Conspicuous amongst these were the famously ferocious Batavians.

In his *Germania*, Tacitus recounted a good deal about this small but influential kindred. They lived on and around the shifting delta of the Rhine, what he called the Rhine Island, 'an uninhabited district on the extremity of the coast of Gaul, and also of a neighbouring island, surrounded by the ocean in front, and by the River Rhine in the rear and on either side'. As the conquests of Caesar and Augustus pushed north into Germania, the military prowess of the Batavians in battles with other natives earned them a unique privilege. Instead of taxes in kind or cash, they supplied soldiers, usually troops of cavalrymen, to the Roman army. There may have been as many as 5000 Batavians in the British invasion force of AD 43, and 40 years later they fought with great ferocity as infantry in the Battle of the Graupian Mountain, also known as Mons Graupius. Probably at the foot of Bennachie, a singular hill in Aberdeenshire, these soldiers from the mouth of the Rhine advanced in the front rank and cut down many Caledonian warriors in Rome's greatest victory in the north. Batavi is cognate to the English word 'better', and so it proved on that bloody day at the Graupian Mountain.

Conspicuous by their battle-hardness and bravery, these Germanic soldiers are much less easy to distinguish through their DNA. The Rhine Island lay immediately to the north of the territory of the Belgae, a people who had migrated to the area of southern England around Winchester. And across millennia, many other groups had either walked from what is now the Low Countries across the plains of Doggerland or sailed the North Sea to make landfall in Britain. It is likely that some Batavian auxiliaries did settle in Britain but their DNA legacy is extremely difficult to identify.

However, they did leave something behind that turned out to be very valuable indeed, and has been described as the greatest archaeological treasure ever found in Britain, something of no intrinsic value but historically priceless. And, as often happens, this treasure was discovered by accident.

In 1970 the brilliant archaeologist, Robin Birley, founded the Vindolanda Trust. Its aims were straightforward: to preserve and excavate the site of the Roman fort that lay on the Stanegate, the Stone Road, approximately halfway between Newcastle upon Tyne and Carlisle. Like the Fosse Way, the road had been laid as a frontier, the northernmost limit of the province of Britannia before the construction of Hadrian's Wall. Birley had no grant or outside help in running the site at Vindolanda and because he depended entirely on entrance fees, he knew that he had to create an attraction. Continued excavation from spring to autumn was both academically valuable and very interesting to visitors. But there was a problem. In order to dig down to the Roman layers in the fort and the vicus on its western edge, Birley's teams had to remove modern field drains. This caused flooding in the long hollow between the remains of the civilian settlement and the walls of the fort and this sometimes prevented access for visitors. The excavators dug a new drainage ditch and to achieve sufficient fall, they went even deeper than before. And ran straight into what looked like a Roman rubbish dump. Having also found the remains of some timber uprights, Robin Birley realised that he had probably come across the earlier wooden fort at Vindolanda, probably the first structure on the site, and he decided to wait and closed up the trenches for the winter.

In the spring of 1973, excavators moved back into the ditch and quickly discovered what seemed like an archaeological layer cake. As the fort was successively rebuilt, work gangs had flattened the foundations of the previous structure and covered them with clay or turf, effectively sealing the archaeology, often in an anaerobic state. As they cut through the layers, the archaeologists brought up all kinds of organic material – even bracken and fern floor covering that quickly began to rot and stink. But one chance find proved to be astonishing, absolutely unique, the greatest treasure of Britain.

In what appeared to be a room of a timber building, rubbish was found, thin slivers of wood that had been preserved in the anaerobic conditions. An excavator noticed that two seemed to be stuck together, and when he pulled them apart, script was visible. Robin Birley recorded the moment: 'I had another look and thought I must have been dreaming, for the marks appeared to be ink writing.' And he later added:

> If I have to spend the rest of my life working in dirty, wet trenches, I doubt whether I shall ever again experience the shock and excitement I felt at my first glimpse of ink hieroglyphics on tiny scraps of wood.

These were the first of what became known as the Vindolanda tablets. Notes, lists and letters thrown away at the fort in the years around AD 100, they revealed aspects of life in Roman Britain that were entirely new, a texture that previously could only have been imagined. One previously uncertain relationship emerged into the light and that was the interactions with the native British farmers and tradesmen who supplied the garrison. Attitudes were, perhaps not surprisingly, very disrespectful. 'Brittunculi' was the dismissive diminutive used. It means something like 'pathetic little Brits'.

The notes and letters have of course much more to say about the lives of the occupying colonists and here is the text of one of the most remarkable of the Vindolanda finds:

Claudia Severa to her Lepidina, greetings. On the third day before the Ides of September, sister, for the day of the celebration of my birthday, I give you a warm invitation to make sure that you come to us, to make the day more enjoyable for me by your arrival . . . Give my greetings to your Cerialis. My Aelius and my little son send him their greetings.

I shall expect you, sister. Farewell, sister, my dearest soul, as I hope to prosper, and hail. To Sulpicia Lepidina, (wife) of Cerialis, from Severa.

It is a timeless letter, something that could have been written almost verbatim on any remote frontier of the British empire 1700 years later. Claudia Severa and Sulpicia Lepidina sound like faintly precious aristocrats trying to make the best of it out in the sticks, their meetings and correspondence islands of civility in a sea of barbarity. But the sheer immediacy, humanity and intimacy of life in Roman Britain is perhaps even better conveyed in this enigmatic exchange between the two:

I, sister, just as I had spoken with you, and promised that I would ask Brocchus, and that I would come to you – I did ask him and he replied that it is always, wholeheartedly, permitted to me, together with . . . to come to you in whatsoever way I can. There are truly certain intimate matters which [I long to discuss with you (?). As soon as I know for sure (?)] you will receive my letter, from which you will know what I am going to do . . . I was . . . and I will remain at Briga. Farewell my dearest sister and my most longed-for soul. To Sulpicia Lepidina from Severa, wife of Brocchus.

These are poignant echoes from the long past. It is easily possible to imagine the breathless meeting between the two, their heads inclined, sitting in a private corner, their voices low, like a conversation behind the fans in a novel by Jane Austen.

What is even more remarkable is the authorship of the Vindolanda letters. Claudia Severa, Sulpicia Lepidina, Aelius Brocchus and Flavius Cerialis were not Romans, or Italians. All four were Batavian aristocrats, people whose barbarian origins were on and around the Rhine Island, whose grandfathers had risen in bloody rebellion against Rome before being reconciled, and whose savagery in battle was the stuff of legend. And yet here they are at windy Vindolanda on the bleak moors of north Northumberland, exchanging pleasantries, behaving like cultured Romans sighing at being so distant from civilisation and each other. In little more than half a century, the Batavians appear to have become thoroughly Romanised. It was the outcome of a deliberate policy. In the *Agricola*, Tacitus wrote fascinatingly about the social changes that were rippling across the new province of Britannia in the second half of the first century AD after the conquest:

> The result was that those who had just lately been rejecting the Roman tongue now conceived a desire for eloquence. Thus even our style of dress came into favour and the toga was everywhere to be seen. Gradually, too, they went astray into the allurements of evil ways, colonnades and warm baths and elegant banquets. The Britons, who had no experience of this, called it 'civilisation', although it was part of their enslavement.

An emperor who spent 11 years in the saddle, touring the frontiers of the empire, Hadrian realised that consolidation was essential. Perpetual conquest simply could not be sustained and limits had to be set. And so he commanded a wall to be built. Even though his successor, Antoninus Pius, pushed the frontier of Britannia further north to the Forth–Clyde isthmus after AD 138, the occupation of southern Scotland lasted little more than a generation.

Writing towards the end of the first century AD, Tacitus offered a rapid Roman review of the British, the people they had conquered and the barbarians who lay beyond Hadrian's and Pius' frontiers. And it turns out that the view down his aristocratic nose was illuminating:

As to what human beings initially inhabited Britain, whether native-born or immigrants, little has been established, as is usually the case with barbarians. Be this as it may, their physical appearance is varied, which allows conclusions to be drawn. For example, in the case of the inhabitants of Caledonia, their red-gold hair and massive limbs proclaim German origin. As for the Silures, their swarthy features and, in most cases, curly hair, and the fact that Spain lies opposite, provide evidence that Iberians of old crossed over and settled this territory. Those nearest to the Gauls also resemble that people. Either their common origin still has some effect or, since the two countries converge from opposite directions, shared climatic conditions produce the same physical appearance.

All the same, it is plausible on a general estimate that the Gauls occupied the adjacent island. You can find their rites and their religious beliefs. The language is not much different, likewise the same boldness in seeking out danger – and, when it comes, the same timidity in facing it. Still, the Britons display more ferocity, having not yet been made soft by prolonged peace. We are told, indeed, that the Gauls, as well, used to be warriors of repute. Then decadence set in, hand in hand with peace: their courage has been lost along with their liberty. The same has happened to the Britons long since conquered. The rest are still like Gauls once were.

At least two of the European connections made by Tacitus chime with the patterns of migrations suggested by DNA testing. The Atlantic route from Iberia may indeed have brought some of the ancestors of the Silures, the powerful kindred who occupied much of south Wales as well as parts of Gloucestershire and Herefordshire. Their name is derived from 'siol', a Celtic root word meaning seed, and Silures probably translates as something like The Kindred, a distant hint of a new, more discrete and different set of lineages from elsewhere. To the west of the Silures lay the lands of the Demetae, a people with personal name and place name links to Atlantic France.

The similarities, cultural and ethnic, between the kindreds of the south-east of England and those on the opposite shore of the Channel will have been obvious to Romans and are fairly well attested in the historical and archaeological records.

It is only Tacitus' statement that the red hair and massive limbs of the Caledonians proclaim German origins that has given historians pause. But DNA can shed a little light.

In 2012 more than 4,000 British people were tested for the red hair gene variants. Of course, no one needs to consult a DNA sequence to discover if they have red hair, all they need is a mirror. What the 2012 research aimed to discover was the percentage of Scots who were carriers of the variant, often unknowingly. And the numbers turned out to be very high, almost certainly the highest percentage per capita in the world. While 6% of all Scots have red hair, more than 36.5% are carriers, 1.9 million out of a nation of 5.3 million.

Redheadedness

All physical colouring is a mixture of two pigments and these are black melanin and red/yellow melanin. In people with red hair, a particular receptor in the pathway for pigmentation, known as MC1R, is disrupted and black melanin is suppressed while red/yellow melanin is allowed to be made. The result is red hair, light skin colour, often freckles and generally a greater sensitivity to sunlight. There are more than forty gene variants that can pass on red hair, but three are most common. Identified by the biochemical that characterises each one, they are: Cysteine-Red (or R151C), Tryptophan-Red (or R160W) and Histidine-Red (or D294H) carried by 2.5% of British people.

For a child to have red hair, both parents must be carriers of one of the variants. There is a 25% chance that their children will have it. This is called recessive inheritance. Everyone who carries one of the variants is a direct descendant of the first person ever to have it. Those with Cysteine-Red have a 70,000-year-old variant that probably arose in west Asia, those with Tryptophan-Red are the descendants of someone who probably also lived in west Asia 70,000 years ago, and finally Histidine-Red belongs to a much younger group which descends from a European who lived about 30,000 years ago.

Now, the reasons for such a high percentage are a matter for conjecture, but they are probably linked to the climate. Intriguingly, Tacitus made a similar connection between the Gauls, the southern British and the influence of a shared climate on appearance. All human beings need to absorb vitamin D from sunshine and skin colour evolves accordingly. Skin darkens where the sun shines often and it lightens where it shines less. Red hair and lighter skin tone are linked and since Scotland's Atlantic climate make it one of the cloudiest countries in the world, the inhabitants have adapted. In Mediterranean Europe red hair is much less common, at a frequency of between 1% and 2%, and if the number of redheads in Caledonia was as high in the first century AD as it is now, the difference will have struck Roman incomers forcibly.

The mention of large limbs and the link with Germany is much less clear but northern Europeans, because of relatively cloudy climates, are more red-headed than the southerners, and it may be that the legions who fought on the Rhine noticed many with red hair in the ranks opposite them. Boudicca was said to have tawny hair, probably a tint of red.

Perhaps sexual selection is a potential reason – that having red hair, although not helping survival, was considered sexually attractive, and thus redheads had more children than non-redheads, leading to an increase in frequency. Like the peacock's tail, it doesn't help him survive but it attracts mates, hence the genes are passed on.

The Roman conquest appears to have made the native British population more mobile. Just as the Sarmatians were posted a long way from their homeland, where they might have made trouble, British regiments of auxiliaries were raised and inscriptions record that they saw active service across the continental empire. Very soon after the invasion of AD 43, the Romans began to raise local levies. It was a simple means of removing potentially trouble-making bands of warriors from a newly conquered province. Two cavalry units and twelve regiments of infantry from Britain were posted mostly to the Danube frontier and one of these won battle honours in the Emperor Trajan's war against the Dacians, the peoples of what is now Romania. And as late as the fourth century, a mysterious kindred known as the Atacotti fought against imperial forces before finding themselves enlisted as Roman soldiers. In a document called the 'Notitia Dignitatum', these warriors are listed on active service. Their name is the only clue to their origins, and it is thought that they either came from the Hebrides or from Ireland. Intriguingly, it means The Old Peoples.

In the second century AD a Greek geographer, Ptolemy, drew the first detailed map of Britain. At first glance it looks as though he made a grotesque and basic error. At about the line of the River Tay, Scotland is suddenly made to bend eastwards so that Caithness extends into the North Sea and points towards the continental coast. In reality it was less of an unintentional blunder and more of a solution to a conceptual problem. Like most classical cartographers, Ptolemy did not believe that human beings, even the shaggy barbarians of the north, could survive in the extreme, un-Mediterranean weather north of 63° latitude. So, instead of extending Britain northwards to an impossible 66°, he bent it.

While the shape of the island is distorted, the detail plotted on the map is very helpful, the first appearance on a record of any sort of many names. Much of England is divided into a patchwork of small kingdoms, but in the north (or rather the north-east), Ptolemy's sources appear to thin out. He names kindreds around the coastal regions of Scotland, perhaps drawing his information from the log of the circumnavigation completed during the period of Agricola's governorship, and perhaps from Pytheas' much earlier records. There is very little sense of what or who might have inhabited the landward regions. But as ever, the names are informative.

Across the far north totem animals were thought powerful: the Lugi of Ross-shire translate as the Raven People, the Orcades were the Boar Kindred, the Caereni of Sutherland the Sheep Folk and the Epidii of Kintyre were the Horse-Masters. In England the names of peoples were different, and they often seem to betray origins. The Belgae are obvious, as are the Parisii of Yorkshire. There were Atrebates in both Britain and Gaul, and a version of the name survives in the city of Arras in the Pas de Calais. Catuvellauni is cognate to Catalauni in Belgic Gaul and it means something like Battle-Leaders, while the Trinovantes of Essex were simply the New People. While these connections veer towards conjectures rather than firm links, when they are taken together with written and genetic evidence, a picture of the south-east of England as Gaul-Across-the-Sea emerges. Foreigners most certainly did not begin at Calais. In recent history the Channel may have been seen as barrier but for many millennia, it was a connector.

A huge land barrier, absolutely intended as such, was built in the north. Hadrian's Wall, the largest Roman monument anywhere, stretching 73 miles from Wallsend, east of Newcastle upon Tyne, to the Solway coast beyond Carlisle, was an emphatic frontier, a clear limit to conquest. Aside from the Antonine interlude in the middle of the second century, the wall divided Rome from the barbarians beyond for almost 300 years, but not always effectively. Dio Cassius sets the scene:

> The two most important tribes of the Britons [in the north] are the Caledonians and the Maeatae; the names of all the tribes have been practically absorbed in these. The Maeatae dwell close to the wall which divides the island into two parts and the Caledonians [dwell] next to them. Each of the two inhabit rugged hills with swamps between, possessing neither walled places nor towns, but living by pastoral pursuits and by hunting.

The wall in question was almost certainly the turf rampart of Antoninus Pius and placenames further help locate the Maeatae around the Ochil Hills and Stirling. Dumyat, a prominent hill over-looking the Forth Valley, derives from the Dun or the Fort of the Maeatae, and near Denny, south of Stirling, Myot Hill rises to the east. These may have marked territorial boundaries. This power-ful kindred were also durable, and the biographer of St Columba, St Adomnan, wrote in the seventh century of a war between the Gaels of Dalriada and the Maeatae – in which his fellow Celts were defeated and their princes killed. The Irish annals of Tigernach reported the same event as the 'battle of Circenn', the name of the later Pictish province of Angus and the Mearns.

Between AD 180 and AD 192, the reign of the increasingly crazy Emperor Commodus, Dio Cassius chronicles more conflict:

His greatest war was in Britain. The tribes in the island crossed the Wall which separated them from the Roman legions, did a great deal of damage, and cut down a general and his troops; so Commodus in alarm sent Ulpius Marcellus against them. Marcellus inflicted a major defeat on the barbarians.

But it was only temporary. The raiding by the northern barbarians, almost certainly the Maeatae and the Caledonians, soon resumed and this time their incursions across the wall were even more serious. They succeeded not only in plundering but also took some high-ranking captives from the Roman province, either soldiers, aristocrats or administrators or all three. By the end of the second century and after one of many periodic bouts of civil war, the great African soldier, Septimius Severus, had made himself emperor but Britain and its problems had to wait. His armies were campaigning in the east. Dio again:

> The Caledonians, instead of honouring their promises, had prepared to defend the Maeatae and Severus was concentrating on the Parthian [Persian] War; so Lupus [governor of Britannia] had no choice but to buy peace from the Maetae for a considerable sum of money, recovering a few captives.

There is a faint hint in this passage of the classic Roman policy of divide and conquer, with a separate peace apparently having been concluded with the Caledonians. Again, respite was short and the treaty made by Virius Lupus was torn up by the Maeatae and Caledonian kings. The new governor was Alfenus Senecio and he will have taken no pleasure in writing this urgent dispatch:

> ... the barbarians had risen and were overrunning the country, carrying off booty and causing great destruction ... for effective defence more troops or the presence of the emperor was necessary.

Enough was enough. Britannia was a prosperous province. Rome had to act decisively and in AD 208 the empire came north. For four years York was the makeshift imperial capital as Severus and

159

his general staff organised a huge force of 40,000 men and a supporting fleet, the Classis Britannica. So vast was the invading army, the largest ever to march into Scotland, that a series of marching camps in Lauderdale have produced fascinating evidence of how slow was their progress north. It took four days for the legions and auxiliaries to march 38 miles from Melrose to Inveresk on the Forth, and when the vanguard had reached the site of the next marching camp, the rearguard was only just passing out of the gates of the previous one. The fleet was waiting, almost certainly having dropped anchor in the Forth, its ships full of supplies for the huge and hungry army. It would shadow the advance as it sailed north up the coastline of eastern Scotland.

Faced with such overwhelming strength, the kings of the Maeatae and the Caledonians sued for peace. But Severus ignored their embassies and his vast army rumbled northwards, bent on punishment, even genocide. In AD 210 a treaty appears to have been concluded after the lands of the Forth and the Tay had been laid waste. There may have been a high-level meeting after terms were agreed. In what sounds like a flight of fancy, Dio reports an encounter between the emperor and the Maeatan king, Argentocoxos. Wives were also present, including Severus' wife Julia Domna, who was notorious in Rome for her affairs. Having stated that the northern barbarians all enjoyed having sex with each other constantly, Dio alleges that the empress teased the queen of the Maeatae with this and she is said to have replied:

We fulfil the demands of nature in a much better way than you Roman women, for we consort openly with the best of men whereas you let yourselves be debauched in private by the vilest.

So there.

A year after this exchange the warbands of the kindreds attacked the legions once more. At York plans were made for a final solution, the annihilation of the Maeatae and the Caledonians, but great bloodshed was prevented by a single death. The warrior-emperor, Septimius Severus, died and his sons hurried back to Rome to contend for his throne.

No doubt there was rejoicing as feasting fires blazed and crackled in the Ochils and the Angus glens. The mighty Maeatae had fought off the empire and its vast army. And then, Adomnan aside, this remarkable barbarian kingdom appeared to fade into the mists of the past. Until now. It seems that the DNA marker of the Maeatae has survived. A Y-chromosome marker, R1b-S190, is carried by about 1% of all Scottish men, around 25,000 in all of today's population. It appears to be closely concentrated in Stirlingshire, Perthshire and Fife.

The descendants of Argentocoxos and his warriors may take deserved delight in their kindred's fame and name. Maeatae means something like The Great Ones. And so they were, great enough to raid into Britannia, take hostages, extract bribes and force the ruler of the greatest empire Europe has seen to marshal a huge invasion force and march north to meet them.

After the upheavals of the early third century AD, Britannia settled once more. Despite a succession of assassinations in Rome and the constant jockeying for imperial power, the province seemed to prosper for 50 years or so. In two separate episodes, Britannia found itself independent, at least nominally. It was part of the Independent Empire of the Gauls between AD 259 and AD 273, a link that may have seemed natural to those who lived in the southeast. And for 10 years the province lay outside the empire, the realm of two usurpers, Carausius and Allectus.

The late third century not only saw political convulsions, raiding also resumed. Both the Picts from north of the wall and the Scots from Ireland made their first entry into the written record. Probably a soldiers' nickname for warriors who wore tattoos, the Picts or Picti seem to be successors of the Maeatae and the Caledonians, while the name of the Scotti had yet to become associated with Scotland. In the east, forts were built on what has since become known as the Saxon Shore. From Brancaster at the mouth of the Wash around the coast to Pevensey on the Channel and beyond, this defensive network may have been manned by Saxon mercenaries of the Roman army employed to repel Saxon invaders. Indeed, there is epigraphic and other evidence all over Roman Britain for the increasing deployment of Germanic units. When Carausius was establishing himself at the end of the third century AD, his field army was largely Germanic and many warriors may well have been promised plots of land in return for military service.

Lying on the farthest edge of the empire, Britannia began to feel the ripples of a European phenomenon. After *c* AD 350, the age of what German historians call *die Völkerwanderung* began. Probably reacting to pressure from the east (although not always), entire peoples began to migrate and they pressed hard on the Rhine and Danube frontiers, often breaking through. These wandering bands fell into three groups: the Scandinavians who later gave rise to the Danish, Norwegian, Swedish and Icelandic communities; the west Germans of the North Sea coast comprising the Batavians, Frisians, Franks, Alamans, Jutes, Angles and Saxons; and the east Germans, who occupied lands east of the Elbe – the Swabians, Lombards, Burgundians, Vandals, Gepids, Alans and Goths. The pressure exerted by the last group caused a tribal traffic jam on the northern European plain, and when the pressure built to intolerable levels, it burst over the western Roman Empire, transforming it and Europe utterly. On the last day of AD 406, to the horror of the frontier forces on the Roman shore, the Rhine froze and more than 70,000 barbarians slithered across the ice to burst into northern Gaul. There was widespread havoc.

In Britannia, the Völkerwanderung's earliest emphatic effect occurred in AD 367. In what is known as the Barbarian Conspiracy, Picts, Scots and Franks attacked in a concerted manner, causing chaos throughout the province. Imperial authority was ultimately re-established by a remarkable soldier, Count Theodosius, but the episode was in reality the beginning of the end for the province. In formal terms Britannia was to last another 40 years, but when the near-contemporary historian Zosimus wrote of the events of AD 408, it seems that links with Rome were finally severed:

The barbarians [from] across the Rhine attacked everywhere with all their power, and brought the inhabitants of Britain and some of the nations of Gaul to the point of revolting from Roman rule and living on their own, no longer obedient to Roman laws. The Britons took up arms and, braving danger for their own independence, freed their cities from the barbarians threatening them; and all Armorica and other provinces of Gaul copied the

British example and freed themselves in the same way, expelling their Roman governors and establishing their own administration as best they could.

When a letter dated AD 410 arrived in Britannia from the chancery of the young Emperor Honorius, it supplied what is often seen as a full stop. It advised the cities of the province to look after their own defences.

Part 2

7

The Wanderers

✻

O N THE NIGHT OF 24 August AD 410, a band of disaffected
slaves gathered in the Gardens of Sallust, not far from the
Salerian Gate inside the most northerly circuit of Rome's
walls. There may have been as many as 300, and they were armed.
And almost certainly well paid for what they were about to do.
Having moved silently through the shadows, they rushed towards
the narrow gateway and its round sentinel towers and overpowered
the guards. Then they quickly slid the bolts and swung open the
double doors. Waiting outside were the ferocious warriors of Alaric,
king of the Visigoths. They swarmed through the Salerian Gate
and raced for the Forum and the centre of the imperial city. For
three days they killed, pillaged and burned. But as Arian Christians,
the Visigoths were surprisingly restrained and discriminating. It
is said that church plate and other valuables were untouched and
consecrated buildings were left intact.

But there was no Roman emperor to take captive. The imperial
administration had recently abandoned its ancient capital. Alaric
had already besieged the city twice before and was only bought off
by vast bribes of gold, silver, silk, pepper and other precious items.
The emperors first chose Milan because it was smaller, safer and
more easily garrisoned and it had served as the centre of the empire
before the young Honorius had decamped his court once more.
They went to Ravenna, defended not only by stout walls but also
surrounded by marshland. Nevertheless, the fall of Rome was a tremendous shock, a tremor that reverberated throughout the empire.
Not for 800 years had an enemy breached the gates; not since the

167

warbands of the Celts had taken the city in 390 BC had Rome been in the hands of hostiles. When the great translator of the Bible, St Jerome, heard the news, he was speechless:

> My voice sticks in my throat, as I dictate, sobs choke my utterance. The city which had taken the whole world was itself taken.

Slowed and weighed down by carts full of plunder and wains of captives, (these included Gallia Placida, the emperor's sister) the Visigothic army moved south to the toe of Italy. As Alaric contemplated an invasion of Sicily and perhaps Africa, he suddenly fell ill with fever and in Calabria he died. Unanimously elected to succeed, Ataulf reversed his brother-in-law's strategy and turned the army back north. In AD 413, the new king seized the cities of Narbonne and Toulouse and established the southern part of the province of Gaul as a Visigothic kingdom. Ataulf did not see Rome as the past but rather as a working political structure that could be given new life – by the very people who had sacked the eternal city and were busily carving up the empire. Here are the words, which seem like a manifesto, put into Ataulf's mouth by the historian Orosius:

> At first I wanted to erase the Roman name and convert all Roman territory into a Gothic Empire: I longed for Romania to become Gothia, and Ataulf to be what Caesar Augustus had been. But long experience has taught me that the ungoverned wildness of the Goths will never submit to laws, and that without law a state is not a state. Therefore I have more prudently chosen the different glory of reviving the Roman name with Gothic vigour, and I hope to be acknowledged by posterity as the initiator of a Roman restoration, since it is impossible for me to alter the character of this Empire.

The speech may have been invented but the sentiments were not. Barbarians were more than impressed by the glory that had been Rome, they were dazzled by it. And also they saw themselves as both legitimised and enhanced by association with it. The empire was not a target to be smashed into smithereens but a template for power, a machine that worked, produced wealth and under barbarian leadership a polity that could prosper once more. All over Europe and in Britain, the destroyers of the empire often aped its culture. And so did those who believed themselves to be its defenders.

Old King Cole is an unlikely historical memory, but it is almost all that remains in popular culture of a shadowy but fascinating figure. Coelius, or Coel Hen in Celtic, was a fragile link between the imperial province with all its prestige and the half-light of post-Roman Britain.

By the middle of the fourth century, the shrinking provincial garrison had been split into three commands. The Comes Britanniarum led a small but mobile field army whose nominal strength was six squadrons of cavalry and three units of infantry, probably no more than 6000 men in all. Its role was reactive, to move quickly to crisis points where frontier forces needed reinforcements to fend off barbarian raids. The Comes Litoris Saxonici, the Count of the Saxon Shore, covered the string of forts on the southern and eastern coasts. And at the old legionary fortress at York, where Septimius Severus had died and Constantine I was proclaimed, the Dux Britanniarum was specifically responsible for the units that garrisoned the forts on Hadrian's Wall. At what became Army Command North, this most senior officer also held overall military command of Britannia.

It may well be that Coelius or Coel Hen was the last Dux, the Duke of the Britains, to be appointed by an emperor in Rome. His name certainly retained the ring of imperial authority and he is often referred to by the Celtic name of Godebog, which means Protector. He may have been the last to lead the eagles, to summon the clash of legionary steel.

In the century after the end of the province, Celtic kingdoms in the north began to emerge from the mists. It was as though, one eminent historian, John Morris, has remarked, the empire had never existed and the thoroughly Celtic nature of these polities reached far back to the Iron Age, to a Britain that existed before Caesar's legions fought their way up the beaches of the Channel coast. Half-forgotten realms like Rheged, Elmet, Gododdin, Strathclyde, Powys, Dyfed, Gwynedd, Ebrauc, Bryneich, Deifr and Dumnonia re-formed themselves, complete with aristocracies, royal dynasties, frontiers and legends of their origins. It may be that this Celtic substructure of Britain always underlay the tidiness of Roman provincial government and that in the north and west, Roman culture had only ever been skin deep. But the lustre of the imperial past was attractive and many of the dynasties of the re-emerging kingdoms claimed descent from Coel Hen, the last Dux Britanniarum, Old King Cole.

The genealogies of royal and aristocratic families can be unreliable. They serve a political as well as a historical function as they seek to legitimise origins and assert rights to rule. An association with the last dux, if Coel Hen did indeed command the soldiers of Rome, could confer the half-forgotten afterglow of imperial dignity. The later pedigrees known as the Bonedd Gwyr y Gogledd, the Descent of the Men of the North, cite Coel Hen as the progenitor of several significant dynasties, so many that his progeny are known as the Coelings. He was said to be the ancestor of Urien, the great king of Rheged, of Guallac of Elmet, the brother-kings of Ebrauc or York, Gwrgi and Peredur, and of Clydno Eiddin, king of Edinburgh.

All of these shadowy realms lay in the north, the part of Britain more directly under the control of the Dux Britanniarum, and also the part that was much less Romanised. Archaeological geography is eloquent on this bias. The principal area of the province where villas were built and estates arranged around them was to the south of a diagonal line drawn approximately from the Humber to the Severn. All of the kingdoms ruled by the Coelings and other native dynasties were to be found to the north and west of that cultural frontier. It may be fairly assumed that native aristocrats had retained their lands and authority, at least to some significant degree, under the Roman provincial radar. And when central imperial control evaporated after AD 410, the northern aristocrats began to call themselves kings, but lest they seem like usurpers, they also claimed Coel Hen as their ancestor, a continuity with the Roman past.

In the south there were many more Roman towns, and archaeology reveals them as centres of government with law courts, a forum, baths, temples and places of public entertainment, all the accoutrements of civilisation. Their inhabitants, some of them Roman citizens, others wealthy, were the people who had most to lose as the empire in the west began to crumble. The production of pottery and coinage, both reliable barometers of economic activity, contracted quickly after AD 410 and towns will have become difficult to sustain as trade was transacted by barter rather than paid for with cash. But what actually happened, at least in any detail, in the Romanised south, a sequence of events, dates and names, is very difficult to establish from the sparse sources.

The principal authority for the period is a polemic, 'On the Ruin and Conquest of Britain', written by Gildas, a Celtic monk with excellent Latin. It is a tirade against the native kings and their fecklessness in the face of barbarian invasions which rarely draws breath sufficiently to note down dates or events. One exception is Gildas' report of what happened in the year AD 443, when the south was hard pressed. Flavius Aetius was to be the last capable Roman general in the west, and in AD 450 he defeated the great horde of the Huns at the Catalaunian Fields in north-eastern Gaul. It was a spectacular and unexpected victory, a last hurrah from a dying empire in its first major province. Gildas quotes an evocative passage in an urgent appeal sent across the Channel, probably from authorities in the beleaguered towns of Roman Britain:

To Aetius, thrice Consul, come the groans of the Britons . . . the barbarians drive us into the sea, and the sea drives us back to the barbarians. Between these, two deadly alternatives confront us, drowning or slaughter.

Two further sources add detail to the fate of fifth-century Britannia. A chronicle compiled in Kent between AD 425 and AD 460 attempted to discover the correct sequence of events. And Bede of Jarrow's *Ecclesiastical History of the English People* offered what has long been accepted as the definitive account of the coming of the Angles, Saxons and Jutes to Britain. Writing in the early eighth century, anxious to legitimise the triumph of the peoples who would become the English and virtually ignoring the native British story of what was unquestionably a tale of two sides, Bede's highly partial account nevertheless forms the basis of the conventional and popular view of how Britannia became England.

Taking all of these sources together, and Bede with a generous pinch of salt (he often follows Gildas in any case), a reasonable chronology for the fifth century can be pieced together. British resistance to barbarian attacks appears to have been effective between AD 410 and AD 442, but this phase was followed by a time of fierce fighting between natives and barbarians attempting to carve out territory and settle. This lasted for about 20 years as parts of eastern England fell under the control of Germanic incomers. The rest of the century saw their advance slowed and even halted.

There were other dynamics to distinguish the two sides. Roman towns had been Christianised and almost all of the kings of the north claimed to march to war with God on their side. Later sources relate that their warriors called themselves Y Bedydd, the Baptised. By contrast, the Angles, Saxons, Jutes and other raiders were pagan, Y Gynt, the Gentiles. Gildas' righteous fury was fuelled by what he saw as the failure of the Baptised to turn back the forces of darkness, 'the vile and unspeakable Saxons, hated of God and man alike'. And unlike the Franks who overran Gaul in the wake of Aetius' assassination in AD 455, these barbarians remained pagan for many generations, until the end of the sixth century. Entertainingly, Gildas also called them 'Furciferes', literally fork-bringers or servants. It is usually translated as rascals.

Celtic terms betray another important contrast. The Welsh word for the Welsh is Y Cymry (the English word is from the Germanic 'Wealhas', meaning slaves). 'Cymry' derives from the Latin 'Combrogi' which means the people who share a common border. The name clearly survives in Cumbria, a difficult, mountainous and still Celtic part of England that resisted the English takeover. But a more extended meaning for Cymry is from 'cives', the citizens. That is probably how the people of the dying province saw themselves, as Roman citizens whose common borders are being assailed by barbarians.

They may have shared a common border but they were not united. Gildas ranted about foolish tyrants unable to combine against a common enemy, and the Kentish chronicle remembered a civil war in the AD 430s. On one side stood the soldiers of Ambrosius Aurelianus 'whose ancestors wore the purple', and whose name survives in Wales as Emrys. On the other side was Vitalinus. He was almost certainly the same man identified by Gildas and later Bede as Vortigern. Here is the entry in the Kentish chronicle:

Then there came three keels [ships], driven into
Germany. In them were the brothers Horsa and H
Vortigern welcomed them ... and handed over to u
island that in their language is called Thanet, in ours Ruoihi.

Vortigern is not a name but a title. It is a version of Vawr
Tigherna, or Overlord, a thoroughly Celtic description of a man
who may have been High King of southern Britain. The reason
he invited the Germanic warriors of Horsa and Hengest is often
forgotten. The Picts of the north had been raiding in Britannia
for at least a century – Gildas called them 'transmarini' – and the
Vortigern or Overlord needed mercenaries to contain them. It has
been characterised as one of the most spectacular misjudgements in
British history. The *Anglo-Saxon Chronicle*, compiled contemporane-
ously by clerics, gives a more complete report of events:

> 449. In this year Mauricius and Valentinian obtained the kingdom
> and reigned seven years. In their days, Hengest and Horsa, invited
> by Vortigern, King of the Britons, came to Britain at a place which
> is called Ypwines fleot [Ebbsfleet] at first to help the Britons, but
> later they fought against them. They then sent to Angeln, ordered
> [them] to send more aid and to be told of the worthlessness of the
> Britons and of the excellence of the land. They then sent them
> more aid. These men came from three nations of Germany: from
> the Old Saxons, from the Angles, [and] from the Jutes.

There were others, no doubt. On the basis of archaeology and
some written evidence, historians have added more widespread
continental origins for these arrivals, but it is safe to say that all
may fall under the heading of Germanic warriors. In any event, the
Vortigern's policy appears at first to have succeeded, since sea-borne
attacks from the Picts quickly fade from the historical record, such
as it is. It was at that point in the mid fifth century, however, that
the Vortigern was said to have made his spectacular misjudgement.
Here is the Kentish chronicle again:

> The king undertook to supply the Saxons with food and cloth-
> ing without fail ... But the barbarians multiplied their numbers,

and the British could not feed them. When they demanded the promised food and clothing, the British said, 'We cannot feed and clothe you, for your numbers are grown. Go away, for we do not need your help.'

They did not go. Instead, this unwillingness or inability to pay would soon come to cost the Vortigern and his people dear. A barbarian rebellion flared in the south-east, and it seems that three battles fought around London and Kent were lost by the British. The Germanic incomers seized more land and began to settle. Bede is clearest on the territory they took:

> From the Jutes are descended the people of Kent and the Isle of Wight and those in the province of the West Saxons opposite the Isle of Wight who are called Jutes to this day. From the Saxons – that is, the country now known as the land of the Old Saxons – came the East, South and West Saxons. And from the Angles – that is, the country known as Angulus, which lies between the province of the Jutes and the Saxons and is said to remain unpopulated to this day – are descended the East and Middle Angles, the Mercians, all the Northumbrian stock (that is, those people living north of the River Humber), and the other English peoples.

Bede's tidy allotment of territory and the precise origins of those who colonised it have been long accepted and, with minimal qualification, it represents the conventional wisdom on the beginnings of English settlement. While the Jutic peoples of Kent and the Isle of Wight have faded somewhat from the map and later perceptions of it, the Saxons were quickly recognised as such by Celtic Britain. In Old Welsh and early Gaelic all of the Germanic incomers were labelled as the 'Sais' or the 'Sassunaich'.

The name of England and its language derive, of course, from the Angles. Their homeland of Angeln, Latinised by Bede as Angulus, meant something like Hookland, probably after the peninsulas of southern Denmark. That particular meaning survives in angling, the more formal term for fishing with hooks.

Given the restricted resources at his disposal, Bede was an

impressive scholar, scrupulous in his sources and an excellent stylist. But his orderly allocation of Jutes, Angles and Saxons is, in historical reality, likely to have been much messier, and more intriguing. In the long war for Britain that followed the reign of the Vortigern, there are clear signs that the traditional ethnic battle-lines between Celtic and Germanic were often blurred.

In the north of England, warriors who were probably Angles from southern Denmark established a base at Bamburgh, on the coast of Northumberland, where a stunning castle now straddles a rocky outcrop. Much of what can be seen of it today was restored in the nineteenth century at the behest of William Armstrong, the Tyneside industrialist. But the name of the castle and the village at the foot of the rock is from Bebba, an Anglian queen, the wife of the warrior-king Aethelfrith. According to the genealogies, he was the grandson of Ida, and in a strangely terse note, Bede recorded that Ida 'began his reign in 547'. It is a bafflingly scant reference to the founder of the ruling dynasty of Northumbria, the glittering kingdom of the seventh and eighth centuries that stimulated such a flowering of art and scholarship, including Bede himself.

The formula 'began his reign' occurs elsewhere in the shadowy stories of the Germanic settlement of Britain, and it may be telling. Ida and his immediate ancestors were probably mercenaries hired to protect the post-Roman kingdom of Bryneich or Bernaccia. It lay a great deal closer to the raiding transmarini of Pictland than the Vortigern's territory of Kent and around the mouth of the Thames. Ida's people may even have been stationed along the northern forts of the Saxon Shore under Coel Hen or his predecessors. In any case, Nennius' *History of the British* adds a little to this scrap of information. This history was compiled some time in the eighth century and it is a fascinating jumble of different sources, what seems like reliable history mixed with stories of giants and dragons. Nennius (probably a north British Celtic monk) wrote that 'Ida joined Din Guauroy to Berneich'. Both placenames are the original Celtic versions of Bamburgh and what is now Northumbria. This is significant. Instead of attaching a new Germanic name such as Wessex or East Anglia, Ida began to reign over Bernicia. This implies a longer association with the territory, rather than an invasion by people who arrived out of the blue, so to speak, and it hints at a Celtic–Germanic fusion. To the south, even fewer scraps of information exist, but the original name of Deifr was retained for the Anglian kingdom of Deira in what became north Yorkshire and Durham.

To the north and south two additional examples reinforce a sense of blurred boundaries. Hailed as the first king of Wessex and the progenitor of its illustrious line of kings, Cerdic bore a clearly Celtic name. Ceretic or Caradog is remembered in Cardiganshire and also in the Latinised name of Caratacus, the British prince defeated by the legions at the Medway. And like Ida, Cerdic 'began to reign' some time in the sixth century, as though he had somehow been waiting in the wings. Amongst his descendants were others with Celtic names, such as Ceawlin, Cedda and Caedwalla, a dynasty apparently clinging on stubbornly to a native British heritage while governing what eventually became the dominant Anglo-Saxon kingdom of Alfred and Athelstan.

To the north the vigorous post-Roman kingdom of the Gododdin was centred on the great castle rock of Edinburgh, Din Eidyn, the Lothians and probably the Tweed Valley. In what was a decisive battle at Catterick in north Yorkshire, around the year 600, the Gwyr y Gogledd, the Men of the North, charged in the van of a Celtic alliance against the ranks of the Angles of Deira and Bernicia. Led by Yrfai map Golistan, Lord of Edinburgh, the Celtic army suffered a crushing defeat and by the 630s the Tweed Valley, and all of the Lothians had fallen under Anglian domination. But Yrfai was not what he seemed. His patronymic of 'map Golistan' translates as the son of Wulfstan, an unequivocally Germanic name. Was he the descendant of another mercenary warlord from across the North Sea? It seems likely. And facing Yrfai's men, in the Anglian host at Catterick, it is almost certain that Celtic kings fought against fellow Britons.

To the west, Tewdrig of Gwent succeeded where Yrfai map Golistan failed. By the shores of the Severn Sea, on the borders of modern Wales, he led his warband to victory over the Saxons some time in the late sixth century. Revered as well as feared, Tewdrig abdicated to lead an exemplary life of prayer as a hermit monk and was later canonised. This half-forgotten Celtic hero had a Germanic name, for Tewdrig is a version of Theodoric. And it was not just any Germanic name, because Theodoric the Great was king of the Ostrogoths, ruler of Italy and a viceroy of the eastern Roman Empire. The tradition of lustrous names from the quasi-imperial past continued with Tewdrig's son, Meurig. His name is derived from Merovig, the founder of the Frankish dynasty that ruled Gaul from the middle of the fifth century. Those who saw themselves as the inheritors of the Roman Empire, no matter how minor, borrowed glory and legitimacy from their European counterparts. And so Bede of Jarrow's battle-lines were far from clear-cut and in the turmoil of the great war for Britain from the sixth to the eighth centuries, Celtic and Germanic leaders probably paid less heed to ethnicity and more to opportunity.

Nevertheless, patterns are discernible and studies of DNA markers and their location and frequency in the modern population supply some clarity about the reach of Anglo-Saxon settlement and Celtic retrenchment. The genetic footprint of the Germanic incomers can be found over much of England, especially in the Midlands, East Anglia, the south-east and the south. Recent research published by the People of the British Isles project described central and south-east England as a real genetic cocktail with traces of the prehistoric population clearly overlaid with those of the Anglo-Saxon and slightly later Danish-Viking settlers. What points up the diversity of these heavily populated parts of England are comparisons with the rest of Britain.

Welsh DNA is significantly different, with some markers such as R1b-S300 not found elsewhere. Cumbria, the north-east of England, Devon and Cornwall and Scotland are also clearly unlike the south. The most genetically distinctive are those who are most distant from the point of Anglo-Saxon penetration. Orcadians have a very significant Scandinavian component in their collective DNA, which is scarcely surprising since the archipelago formed part of the Norwegian kingdom until the fifteenth century.

Perhaps the most striking example of history and the study of DNA markers informing each other at this poorly chronicled time in our history is to be found in the south-west of England and Wales. People from Cornwall, Devon and Wales are undoubtedly genetically different from the rest of Britain. Scientists have conjectured that their DNA more resembles that of the prehistoric, pre-Roman population and that this was reinforced as Anglo-Saxon invaders pushed ever further westwards in the sixth century. The Battle of Dyrrham in 577 was a turning point, and after an emphatic victory, the warbands of Wessex captured the crumbling Roman towns of Gloucester and Bath and reached the Severn Sea. But it appears that they did not venture beyond the River Tamar.

Hwicce

After the pivotal Battle of Dyrrham or Deorham in 577, the kingdom of the Hwicce was established over much of what is now Worcestershire, part of Warwickshire and much of Gloucestershire. It included the post-Roman town of Viriconium and the diocese of the Bishop of the Hwicce came into being after 679/680. It in turn became the diocese of Worcester. It was at least as valuable a kingdom as that of Essex or Sussex. Some historians have linked the name to witches and what in present times is called the cult of Wicca. But since it can also mean an 'ark, chest or locker', Hwicce seems more likely to be a reference to the flat-bottomed and fertile valley that lay between the Cotswolds and the Malvern Hills, the heart of this little-understood Dark Ages kingdom.

What did happen was a mass migration across the English Channel. The Roman region of Armorica in north-west Gaul became known as Little Britain, or Brittany. In the fourth, fifth and sixth centuries, as the Anglo-Saxons advanced, large numbers fled south by ship and districts of Armorica became known as Cornouaille after Cornwall and Domnonea after Devon, and settlements such as Bretteville still recall the arrival of British refugees. Certainly there are clear links between modern Y-chromosome DNA markers in the Celtic south-west and Brittany. Some fled even further, to Galicia in the north-west of the Iberian peninsula. Towards the end of the sixth century a bishop was consecrated for the specific ministry of caring for the exiled British population.

The Deepeners

The kingdom of Dumnonia endured long after the collapse of the Roman province of Britannia, until the eighth century. Centred on Devon, Cornwall and parts of Somerset and Dorset, its name probably comes from a proto-Celtic root 'dubno', meaning both 'deep' and 'the world'. Its inhabitants did indeed deepen the earth as they mined for tin. The cognate name of Damnonia was also applied to the Upper Clyde and North Lanarkshire, where coal may have been mined from exposed outcrops or heughs. In Old Welsh, the language spoken in Dumnonia before the eighth century, and in what became Cornish, it was called Dyfneint, or Devon. In Cornwall, there was a kindred named the Cornovii who gave their name to the county. The old kingdom remained firmly Christian after the end of Rome and it is patterned by many obscure and often unique dedications such as St Piran, St Petroc and St Keyne. Two waves of migration across the Channel to Armorica (which became Brittany) left Dumnonia's shores, and for a time kings may have ruled at the same time on both sides of the water. Shadows of the independence of the ancient kingdom continued into the medieval period with the Stannaries, a form of limited self-government underpinned by the importance of the tin trade.

Elsewhere in the fifth century, those who preyed upon the dying Roman province of Britannia were expelled. In a fascinating episode, it seems that from the north, the kingdom of Manau-Gododdin, an expedition rode south. Here is the text from Nennius' compilation:

Cunedag, ancestor of Mailcunus [Maelgwn] came with his eight sons from the north, from the district called Manan Guotodin, 146 years before Mailcunus reigned, and expelled the Irish ... with enormous slaughter, so that they never came back to live there again.

Like Vortigern, Cunedag or Cunedda means something like 'Great General' or 'Great Leader', and the expedition appears to have mustered in Manau, a place that can be located on a modern map. What used to be Scotland's smallest county was Clackmannanshire, and still standing in the old county town of Clackmannan is the Clach na Manau, the Stone of Manau. It lies in the shadow of the Ochil Hills, the southern limits of early Pictland, and it may be that Cunedda and his warriors were summoned or employed by the Vortigern to deal with incursions into Britannia by barbarians from the west.

The name of the Lleyn Peninsula in north Wales is cognate with Leinster across the Irish Sea, because for several generations it had been settled by Irish colonists. Tradition holds that Cunedda expelled them 'with enormous slaughter' and went on to establish himself as the progenitor of the kings of Gwynedd. Shrouded in the mists of uncertainty, half-guessed at, this episode may show attempts to preserve the legacy of Rome and its province and to assert a separate Celtic identity at the same time. If it appears myth-historical, the expedition of Cunedda is no more or less reliable than the arrival of Hengest and Horsa, names that mean Stallion and Horse. The sole difference is that one is reported in Bede's *Ecclesiastical History of the English People* and the other is not.

Irish emigration to Britain in the late fourth and fifth centuries is lent some substance by a fascinating study of a particular DNA marker. In 2006 geneticists at Trinity College Dublin described a lineage later identified as R1b-M222 and their sampling showed that this Y-chromosome marker was very common indeed in Ireland. Its distribution turned out to be heavily weighted to the north, with 40% of all Ulstermen carrying it, 30% in Connaught and 10% to 15% in Leinster and Munster. It is also found in 7% of all Scotsmen but only 1.8% of Englishmen. M222 has a story to tell.

It appears that all of these men, about 1.6 million in Britain and Ireland, and many more emigrants now living in North America and elsewhere, share a recent common ancestor. In the past, powerful men were in the habit of having sex with many different women as well as enthusiastically attempting to father as many legitimate sons as possible to ensure the continuation of their line. This process is coyly known as social selection, and its most spectacular expression is to be found in central Asia where a staggering number, 16 million men, all descend from a single individual, the great conqueror Genghis Khan. In the more modest case of M222, it seems that the most likely progenitor for this great host of direct Irish and British descendants was Niall Noigiallach, the first who could claim to be High King of all Ireland and the ancestor of the great Uí Néill kindred. They dominated Ireland and the high kingship from the sixth to the tenth centuries.

Niall's epithet of Noigiallach means 'of the Nine Hostages', and it boasts of the reach of the High King. To ensure obedience and good behaviour, nine sub-kings were required to send hostages, often members of their family, to Niall's court. Much else about the reign of this remarkably fecund man is ahistorical. Real events, places and people are only very occasionally glimpsed in the legends, genealogies and early fabrications of annals whose authors strain to reach back into the darkness of the past for a bogus completeness. But it appears that Niall was overlord of the five provinces of Ireland: Ulster, Connaught, Leinster, Munster and Mide or Meath; and of four kingdoms across the Irish Sea. One was reputed to be in what is now Scotland, another in Britannia, and two more amongst the Saxons and the Franks, possibly from northern France. The distribution of the DNA marker of M222 does appear to confirm Niall's reach across the Irish Sea, or more precisely, across the North Channel.

Speakers of Scots Gaelic occasionally find it irksome to be reminded that their beautiful language is in fact a dialect of Irish Gaelic, and until the eighteenth century it was often referred to as Irish or Erse. Once again, placenames hint at stories and Argyll, the rugged coastline of peninsulas, headlands and islands of south-west Scotland that faces Ireland, is an anglicised version of Earraghaidheal. It means the Coast of the Gaels, that is, the Irish Gaels. From the fourth to the sixth centuries, and perhaps later, Gaelic-speaking warbands crossed the North Channel and colonised Argyll – as one chronicler put it, 'the sea foamed with the beat of hostile oars'. A remarkable document of the mid seventh century, the Senchus Fer nAlban, lists three kindreds on the Coast of the Gaels: one on Islay, a second in Lorne and Appin and a third in Kintyre. Their kings would one day come to rule over all Scotland. And the DNA they brought from Ireland, from Niall Noigiallach and his many sons, is very significantly present on western coasts and in the Hebrides. M222 was also carried by Irish warbands to the Isle of Man and to a lesser extent to Wales, but it is rarely found in the Midlands and the south-east of England.

Towards the end of the fifth century the westward drive of Anglo-Saxon settlers appears to have been contained, even halted. Giving a rare date and a nugget of tantalising detail, Gildas reported the defeat of a Germanic army at Mount Badon in the year 500. Nennius adds a great deal more. Probably drawing on a praise poem (the remnants of metre and a rhyme-scheme can be detected) composed after a successful campaign, he concludes with a victory at Badon. The British armies were led not by a king but by a warlord, the Dux Bellorum, and his name was Arthur. This famous passage spooks many historians, but there is nevertheless a consensus that an Arthur-like figure did campaign against the hated Sais and was victorious. Mount Badon was probably somewhere in the West Country and Gildas believed that it ushered in a period of relative peace and consolidation, what he called 'our present security'.

Numbers are notoriously difficult to estimate, but even if the overwhelmingly male population of Germanic incomers was only 10% of the total in Britain around 500, scientists reckon that within the span of only five generations, it could have risen as high as 50%. Anglo-Saxon warriors who dominated British society in the east of England and took native wives could quickly outbreed indigenous males, although it is clear that they did not entirely extinguish their DNA signal.

The brief lives of two forgotten English kingdoms, Lindsey and Elmet, offer a sense of this pause in Britain's history after the respite gained by victory at the Battle of Mount Badon – but it should be borne in mind that that is a judgement based on hindsight and not a perception that would have been recognised by contemporaries. England could well have remained culturally and genetically divided into clear eastern and western zones, and these transient polities may have survived and their vanished boundaries become permanent.

In the fifth century there were reports of widespread flooding around the shores of the North Sea. Sea levels were apparently rising significantly and may have had the effect of inundating some coastal settlements in Denmark and Saxony, and also making it easier for those driven west to find new land to penetrate deeper into the interior of England. The Anglian kindred known as the Lindisfaras or Lindisfarena, seem to have sailed up the Humber to settle in Lindsey, the part of north Lincolnshire that took their name. Or perhaps it was the other way around: Lindsey means 'the island of Lincoln' and Lindum was the Roman name of the veterans' colony established there in the first century. And it was almost an island. The rise in sea levels had created wide areas of wetland around the Humber and the Wash that isolated higher ground.

The Foss Dyke (different from the Fosse Way) marked the southern frontier of the kingdom of Lindsey and Roman Lindum appears to have been its capital, the place from which a well-attested line of kings ruled. Ignoring the early flourishes of the genealogies where clearly mythic names such as Woden had been written in, the first to reign in Lindsey was Winta. His name is remembered in two villages north of Scunthorpe: Winteringham means the homestead of Winta's people and Winterton is the farm of Winta's people. These may well have been amongst the first coastal settlements of Anglian incomers to Lindsey. By *c* 500, this small kingdom seemed to be sufficiently powerful to attract the attention of Arthur, or the British warlord who won the battle at Mount Badon. Lindsey was also known as Linnuis, and Nennius records that three battles were fought between Germanic soldiers and the armies of the British 'beyond another river which is called Dubglas'. While the kings of Lindsey survived these assaults, they did eventually fall under the hegemony of the Northumbrians. And after maintaining a degree of independence in the seventh century, the wetlanders south of the Humber were forced to recognise the Mercian kings as their overlords.

Poets remembered Elmet. To the north-west of Lindsey, at the foot of the Pennine range and across what is now the West Riding of Yorkshire, it remained independent until the early seventh century. Its kings were praised by the Chief of Bards, Taliesin Ben Beirdd, who sang of the prowess of their warriors in the halls of the west and the north. His voice echoed down the centuries and lit the imagination of Ted Hughes, who spent his early childhood in west Yorkshire. With the photographer Fay Godwin, he published *Remains of Elmet* in 1979 and in its preface he wrote of history, change and loss:

183

The Calder valley, west of Halifax, was the last ditch of Elmet, the last British kingdom to fall to the Angles. For centuries it was considered a more or less uninhabitable wilderness, a notorious refuge for criminals, a hide-out for refugees. Then in the early 1800s it became the cradle for the Industrial Revolution in textiles, and the upper Calder became 'the hardest-worked river in England'. Throughout my lifetime, since 1930, I have watched the mills of the region and their attendant chapels die. Within the last fifteen years the end has come. They are now virtually dead, and the population of the valley and the hillsides, so rooted for so long, is changing rapidly.

Placenames also recall Elmet. To the east of Leeds lie Barwick-in-Elmet, Scholes-in-Elmet and Sherburn-in-Elmet, and the parliamentary constituency is now known as Elmet and Rothwell.

The Tribal Hideage

Compiled some time between the seventh and ninth centuries in Anglo-Saxon England, this remarkable document lists the names and approximate territory of 35 tribes or kindreds. Almost exclusively dealing with the peoples who lived south of the Humber, the list is headed by Mercia and mostly consists of its neighbours. The Hideage measures wealth or 'hides' of land, and such as the kingdom of the East Angles in Norfolk and Suffolk had 30,000 hides, as did Mercia. It is thought that the list was compiled to allow the powerful kings of Mercia to assess how much tribute their subjugated neighbours could afford. But one of the chief interests of the Tribal Hideage is the wealth of names. Some are entirely obscure – the Herefinna, Noxgaga, Ohtgaga and the Hendrica. Others, like the Hwicce, can be found on the faded map of Anglo-Saxon England only through placename evidence. The Lindisfarena of Lindsey were said to hold 7,000 hides of land while the Elmetians only had 600. Some scholars believe these figures to be wildly wrong but the list of names hints at a pattern of settlement that was well understood at the time and may have betrayed earlier European origins.

This is a powerful, surprising continuity. For, as Ted Hughes understood, history has changed the landscape radically, and industry has covered over the marks of the past before being itself effaced. Perhaps Elmet remained lodged in the poetic and popular imagination because it lasted longer, a Celtic remnant in what might now be seen as Anglo-Saxon/Danish England. It lay well to the south and east of the more recognisably native British cultures of Cumbria and Wales. Its kings ruled over fertile and valuable flatlands rather than being forced to seek the fastnesses of the hill country of Britain where they could more securely guard their power and preserve their culture and language.

Elmet probably emerged from the uncertainties of the former Roman province in the fifth century, perhaps carved out of the patrimony of Coel Hen, the last duke of the Britains. He may have issued orders from the legionary fortress at York, whose massive defences stood long after the remnants of the VI Legion departed. From *c* AD 460 to 616 or 624, five Elmetian kings reigned and perhaps the most far-famed was Guallauc. According to Nennius, he led the armies of Elmet north to join a coalition of British Celtic kings under the command of Urien of Rheged, whose kingdom stretched from the farthest west of Galloway, hinged on Carlisle, and reached south, perhaps down into what is now Lancashire. Those who owed allegiance to Urien may have bordered the lands of those bound to the kings of Elmet.

Having harried the Anglian warbands of Ida's descendants and captured their stronghold of Bamburgh, the British kings drove them to seek refuge on Ynys Metcaud, the tidal island of Lindisfarne, off the Northumbrian coast. The allies camped on the mainland opposite and laid a siege. Victory seemed assured, for when food ran short, the Angles had nowhere to go but the beaches and their boats. But furious dissent flared amongst the British host. Morcant Bwlc, King of Bryneich, is said to have been jealous of Urien's renown and had him treacherously murdered in his tent. The coalition disintegrated, the Anglian warbands crossed the causeway and broke out of the siege, and history took a different turn.

After Bryneich was transformed into Bernicia and the great warrior-king Aethelfrith welded it to Deira in the south, the juggernaut of Anglian power began to roll across the north. Exhorted by their wizards and roared on by their chiefs, the hordes of the Gentiles, Y Gynt, scattered the cavalry of Y Bedydd, the Baptised, at the pivotal battle at Catterick in 600. And Christian Elmet was endangered.

West Yorkshire placenames recall that fragile early piety. Eccles is a version of the Latin 'ecclesia', meaning a church, and more particularly in early Christian Britain, a central church or mother-church, a religious focus for a wide area. Eccleshall, Ecclesfield and several other examples are to be found inside the ancient bounds of Elmet and they hint at a well-organised and widespread faith. And Bede knew of a monastery in the Forest of Elmet that may have been long established. Christian burials of the period are usually recognised by the *hic iacet* formula, and as far west as Gwynedd in Wales an inscription that read 'Aliotus Elmetiacos hic iacet' was found. This and the fact that a district of Dyfed was known as Elfed, the Welsh rendition of Elmet, suggests that there may have been a westward migration.

This was probably prompted by an Anglian invasion in either 616 or 624. Led by King Edwin of Northumbria, the warbands overran Elmet, and around Barwick there is some slight evidence of a last ditch. Around a fort still visible in the village, Ceretic ap Guallauc, the last king, may have had a defensive ditch dug. But it failed to keep out the invaders and after the early seventh century Elmet faded from the political history of Britain. With its fall, most of lowland England came under the control of Anglo-Saxon kings. But Elmet did not fade from the imagination of poets.

Britain was beginning to take shape. Boundaries were forming between the Celtic north and west and the rest. Not yet known as such, England was slowly expanding to its modern extent. The conqueror of Elmet, Edwin of Northumbria, arrogated to himself the honorific title of Bretwalda, Britain-Ruler, the overlord of the old Roman province of Britannia. And he basked in the after-glow of ancient prestige. Whenever the Bretwalda and his retinue progressed around the estates of the kingdom to consume food-rents and make use of rights and services, the trappings of imperial Rome preceded them. A standard-bearer carried a tufa, a winged orb, and the royal household and warband was known as the comitatus. Bede recorded that Edwin's armies pushed far to the west into Wales and in 629 laid siege to the sacred island of Anglesey. By the last quarter of the seventh century, the Welsh authors of the *Brut y Tywysogyon*, the' Chronicle of the Princes', rang a doleful note after hearing of the death of Cadwallon, King of Powys, 'And from that time onwards the Britons lost the crown of the kingdom and the Saxons won it.'

Mercia, midland England, had grown into the most powerful of the Anglo-Saxon kingdoms by the middle of the eighth century, after Northumbria had faltered, and it too began to expand westwards. Settlers moved into the valleys of the rivers Dee, Wye and Severn on the edges of what would become Wales. Cadwallon's kingdom of Powys had once extended into Cheshire and beyond but much territory was lost and it shrank back into its upland heartlands. As early as *c* 550, a kindred known as the Wreoconsaete had occupied parts of eastern Powys. They have left their name at the Wrekin, a singular hill in the Shropshire Plain and the site of an ancient hillfort, and in the town of Wrexham. But the patchy and imprecise picture of the Anglo-Saxon takeover is well illustrated by the story of Viroconium, near Shrewsbury.

First established as a legionary base for the XIV Gemina in AD 58 in preparation for the invasion of Wales and the famous attack on the Druidic shrines on Anglesey, it grew into the fourth largest Roman town in Britannia, with a population of approximately 15,000. But in contrast to the towns, villas and estates in the south and east, Viroconium continued its urban life after AD 410 and the formal severance of ties with Rome. In fact, between 530 and 570 there appears to have been a substantial amount of rebuilding on its 173-acre site. The old stone-built town hall or basilica was carefully demolished and replaced with a series of timber buildings erected on rubble bases. Thirty-three new structures were designed and skilfully built by experienced craftsmen who knew how to use a Roman system of measurement.

Viroconium is a remarkable and enigmatic survival. Historians have conjectured that the work was directed by a Christian bishop based in the town and repairs and renewals appear to have been carried out over a long period, for perhaps 75 years after 570. And then at the end of the seventh century, the town seems to have been peacefully abandoned, presumably because the population had shrunk drastically.

The sun had long set on the Roman Empire in the west, but this vigorous memorial to an urban past appears to have lasted longer than any other settlement in Britain except for one. The fabric of Roman Carlisle was still the focus of some wonderment as late as 685 after the Northumbrian kings had swept aside Urien's dynasty in Rheged. Bede recorded a royal visit made in the company of the greatest saint of the north:

> Cuthbert, leaning on his staff, was listening to Wagga the Reeve of Carlisle explaining to the queen the Roman wall of the city ... the citizens conducted him around the city walls to see a remarkable Roman fountain that was built into them.

'Citizens' or cives was the term used by Bede, and he was usually precise. People still lived in the old city and a system of supplying water, with all the necessary piping and perhaps even an acqueduct, was apparently still being maintained at the end of the seventh century. The walls enclosed a Roman street grid which continued to be respected. After 698 the recently sanctified Cuthbert was honoured with a church built and dedicated to him and the important east to west alignment typical of early Christian churches was altered so that the façade was square on to an existing Roman street. As late as the twelfth century remnants of classical architecture in Carlisle were noted by William of Malmesbury, who was impressed by an arch that carried an inscription to Mars and Venus. The medieval chronicler was also deeply impressed with the surviving paved streets.

What both Viroconium and Carlisle required was an agricultural hinterland to sustain their populations. Markets needed to be held regularly and some method of equitable exchange agreed. This was probably based on barter rather than coinage. It is difficult to see where the latter may have originated. Mints ceased production in the late fourth century and after AD 410 no consignments of coins were imported to pay soldiers or imperial officials. What these two relics of urban Roman culture imply is not conflict between incoming Anglo-Saxons and the existing Romano-British town-dwellers, but some degree of cooperation. Perhaps it suited the aspirations of the leaders of the Wreoconsaete to be seen walking in the streets of a Roman town or visiting the basilica. Or perhaps the Bishop of Viroconium continued a late imperial tradition and was the employer of the incoming warband, using their services to protect his town and diocese from their Germanic cousins to the east or from the native kings of Powys to the west.

Viroconium lay in the heart of fertile flatlands, land lost by the Powysian kings when they had been driven back into the Welsh hills. In 655 and again in the early eighth century, native expeditions raided into the territory of the Wreoconsaete. They must have caused considerable damage and disruption, for the Mercian response was spectacular – and is still visible.

Ditches and their upcast had been used to demarcate bits of Britain for millennia, but Offa's Dyke is the longest and most famous such frontier, in fact the longest such earthwork in Europe. King of Mercia from 757 until his death in battle at Rhuddlan against the Welsh in 796, Offa fought fierce campaigns against the Welsh princes, and to leave no doubt where their territory ended and his began, he commanded a 'vallum magnum' to be dug from sea to sea. Stretching 220 kilometres, it runs from the Dee estuary in the north to the Severn in the south. Imposing, vast and emphatic, the dyke is nevertheless poorly understood. It was never garrisoned and could never have been defended. Perhaps it was principally intended as an obstruction. Its mass and height made it very difficult for horse-riding Welsh warbands to cross quickly, and near impossible for them to raid cattle and drive them back westwards when speed was even more important. Like Hadrian's Wall, it may also have been a demonstration of power, work on a grand scale ordered by a great king in possession of immense resources. Offa called on his people, presumably those living in the west of his kingdom, to give military service and work in gangs to dig a 2-metre ditch and form a 7-metre rampart to the east of it. The whole layout is more than 20 metres across. It is a tremendous testament to the importance of clarity about the ownership of land and one of Britain's greatest cultural as well as political frontiers. The modern borders of Wales run close by.

But at the time of its making, the great earthwork was not a precise boundary. In the eighth century there were English speakers living to the west of it and many native communities to the east. In the making of Anglo-Saxon England in the century before, not only were boundaries pushed and raids made and repelled, there was also occasional slaughter. After the Battle of Chester in 616, Aethelfrith ordered the massacre of 1200 monks from Bangor Is-Coed because they had opposed him with their prayers, according to Bede's unblushing report. But such grisly events seem to have been rare. Acculturation was more common, forced or otherwise. The culture and language of the native British to the east of Offa's Dyke survived for some time after its ramparts were thrown up. When the Laws of Ine, King of Wessex, were codified some time around 690, eight clauses dealt specifically with his native British subjects and their rights, and these remained in force for many generations.

In contrast with the fate of the provinces of the western Roman Empire, where invading barbarians overran them quickly (by AD 486, Gaul was entirely in the hands of the Franks) and where native Latin-based languages were rapidly adopted by the conquerors, the Anglo-Saxons took a long time to extend their rule over what is now England. More than three centuries elapsed between the first landings in Kent and the making of Offa's Dyke – and even then not all of the old Roman province was subjugated. In all that time, it seems that differences entrenched as ethnic and racial consciousness deepened on both sides.

Language exchange is instructive. While the Celtic speakers of Roman Britain borrowed freely from Latin – there are more than 600 recognisable loan-words that have survived in modern Welsh – English took almost nothing from native British speech. Instead, those who found themselves serving Anglo-Saxon masters were forced to learn the new language of subjection. Beliefs may also have kept the two cultures apart. The incomers remained pagan for a century or more after the first contacts and the natives, the Baptised, had been mostly Christian since the later Roman Empire and Constantine's edict. The land also took on the marks of change. Some Celtic placenames did survive but most were replaced by English. Perhaps the most poignant of all is the Welsh word for England. It is Lloegyr, and it means the Lost Lands. But intriguingly, its precise meaning attaches more closely to the Romanised region of England, that is, east of a line drawn from the Humber to the Severn, but excluding Devon and Cornwall. Perhaps it can be said that the Celtic nature of Elmet, Rheged, Powys and Dumnonia was never completely submerged.

8

The Sons of Death

✵

Wee, sleekit, cow'rin, tim'rous beastie,
O, what a panic's in thy breastie!
Thou need na start awa sae hasty,
Wi' bickering brattle!
I wad be laith to rin an' chase thee,
Wi' murd'ring pattle!

I'm truly sorry man's dominion,
Has broken nature's social union,
An' justifies that ill opinion,
Which makes thee startle
At me, thy poor, earth-born companion,
An' fellow mortal!

THE FIRST STANZAS OF Robert Burns' 'To A Mouse, On Turning Her Up In Her Nest With The Plough' remember an ancient bond, one that scientists have discovered is much closer than even one of the world's greatest poets could imagine. When he wrote so spontaneously of the startled, scampering mouse as his earth-born companion, Burns may have had the nocturnal scratchings in the thatch and dark corners of his cottage in mind, but he could not have known how the movements of mice and men across the world turned out to be so intimately linked by studies of their DNA.

'Commensal' is the word used by scientists to describe this intimate relationship, and it literally means eating from the same table. Wherever people moved, mice found a way to move with them, and one of the more unlikely examples of nature's social union is to be found in the blood-soaked and spray-washed story of the Viking attacks on Britain and the subsequent settlement of the islands of the North Atlantic. Cowering and timorous under sacks of meal or grain, or hiding under sailing gear or amongst animal forage, Norwegian mice crossed the North Sea on the open decks of the longships of the sea-lords, known as drekar, dragon-ships, and the less sleek knorrs of the colonists and traders who followed them.

When scientists analysed the mitochondrial DNA of house mice on Orkney, they discovered that it differed from that of their southern British cousins. Instead, it matched the mtDNA of Norwegian mice, and since these tiny creatures are unlikely to have swum the North Sea, they must have stowed away in the ninth century, when the Vikings began to raid the Northern Isles and eventually settle there. Native British mice were found to be genetically related to German mice, and while some will have crossed in the keels of the Anglo-Saxons, it is likely that many more skittered across the watery plains of Doggerland before 4000 BC to be eventually turned up by Robert Burns' plough.

More recent studies have shown that the mice of Iceland are also descended from Norwegian ancestors. These mice may have made their way north in the ninth century, travelling via Orkney, Shetland or the Hebrides with a group of sea-lords led by Aud the Deep-Minded, who may have been among Iceland's earliest colonists. Scientists from York University extracted mtDNA from mouse skeletons that were 1,000 years old, dating back to the early arrivals on Iceland in the late ninth century. They then compared it to the DNA of modern mice found in nine locations around the island and saw that it was an exact match. Because of its relative remoteness, Iceland has had too few significant mouse invasions to alter the original genetic signature. More research tracked the migration of mice across the Denmark Strait to Greenland, but when the colony of people faltered at the beginning of the Little Ice Age and then failed, so did their fellow-mortal rodent companions. And when settlers later returned to Greenland, the mice came back with them. In Newfoundland around L'Anse aux Meadows, where Vikings briefly settled before 1000 AD, no traces of Viking mice have been found: the colony was too fleeting.

Arctic Vikings

Excavations at Tanfield Valley at the southern end of Baffin Island by archaeologists from Aberdeen University have confirmed that Viking colonists lived at latitudes inside the Arctic Circle. Whale-bone fragments pierced by drilled holes confirm these pioneering settlements. The native Inuit of the Dorset culture did not use drills but made holes by gouging. The outlines of sod-built houses, not unlike those at L'Anse aux Meadows in Newfoundland, have been identified and rat fur, from black rats, also found. The latter is an Old World species that must have found its way to Baffin Island by ship. The sheltered cove and grassy plateau of Tanfield Valley is likely to have been a summer settlement, a base for hunting and fur trading with the Inuit. When the winter began to close in, the small colony will have set sail for the Greenland villages. Three other sites inside the Arctic Circle have been identified as Viking settlements and one lies on Ellesmere Island, halfway between Hudson Bay and the North Pole.

By contrast, Viking men made an immense impression on the story of Britain. Here is an unusually long and florid entry in the laconic lists of the *Anglo-Saxon Chronicle*:

793. In this year dire portents appeared over Northumbria and sorely frightened the people. They consisted of immense whirlwinds and flashes of lightning and fiery dragons were seen flying in the air. A great famine immediately followed these signs, and a little later in the same year, on 8th June, the ravages of the heathen men miserably destroyed God's church on Lindisfarne with plunder and slaughter.

A shudder was felt through the Christian west. The shrine of St Cuthbert had been desecrated by savages who seemed to sail out of nowhere. At the court of Charlemagne, a magnet for scholars and intellectuals, Alcuin of York held an influential role as an advisor to the great soldier-king attempting to re-form the old provinces of the Roman Empire. Horrified at the news from Northumbria, he wrote no fewer than seven letters to the Archbishop of Canterbury and three to Aethelhard, the bishop who had failed to protect one of the most sacred churches in northern Europe. Just as Gildas had more than two centuries before, Alcuin saw the attack as God's punishment, in this case for monastic backsliding. But amongst all the literary formalities, there is no disguising his genuine outrage as he goes back into Anglo-Saxon history, missing all sorts of ironies:

> It has been nearly three hundred and fifty years that we and our fathers have lived in this most beautiful land . . . Never before has such a terror appeared and never was such a landing from the sea thought possible . . . the church of St Cuthbert [is] spattered with the blood of the priests of God.

Lindisfarne was probably not the first landfall made by the raiders who came to be known as Vikings. Offa of Mercia had given orders in 792 for the coastal defences of Kent to be strengthened, and in 789, the *Anglo-Saxon Chronicle* reports:

> . . . there came for the first time three ships of Northmen. The Reeve rode out to meet them and tried to force them to go to the king's residence, for he did not know what they were; and they slew him. These were the first ships of Danish men which came to the land of the English.

The attack on Lindisfarne four years later had a profound impact and it was followed by raids on other churches where valuable and portable sacred objects were cared for by monks and priests, people unlikely to offer much resistance. Iona suffered particularly badly, with Viking sea-lords plundering and killing in 798 and 802. In 806 they slaughtered 68 members of the community, both monks and lay brothers. Abbot Cellach had little option but to make preparations to abandon the beautiful but vulnerable little island made sacred by the exemplary life of St Columba. A later monastic chronicler wrote that the Vikings were the Sons of Death.

Contemporaries used other names that were perhaps more informative, and hinted at the origins of all this elemental savagery.

'Lochlannaich' were the inhabitants and Lochlann was how the Gaelic speakers of Scotland and Ireland described Scandinavia, and more specifically, Norway. And it seems to mean nothing more complicated than the Land of the Sea Lochs, the fjords. It eventually found its way into the canon of Irish and Scottish surnames as McLoughlin, McLaughlin and MacLachlan as well as the Christian name of Lachlan. In Welsh, Vikings were known by a variant, Llychlyn. More specific, and more confusing, the *Anglo-Saxon Chronicle*'s entry for the landing in Dorset in 789 believed that three boatloads of murderous Danes had arrived, but another version of the same incident recorded that they came from Horthaland in western Norway.

History rather than geography attached other names. In the early ninth century, as raiding intensified around the coasts of northern Britain and Ireland, chroniclers took to using a Latin term that Gildas had applied to the Anglo-Saxons. With their total disregard for the sanctity of churches, the Vikings were pagans or genti, gentiles, and in Welsh, Y Gynt. Gaelic speakers also knew them as Gall, or foreigners. Deriving from the Latin word 'Gallia' for what is now France, the reason for the adoption of this name is now lost in the mist and low cloud of early medieval history. But it is still used. Non-Gaels (that is, non-Gaelic speakers) in Scotland and beyond are the Gall, but paradoxically, the Gaelic heartland of the Hebrides is known as 'Innse Gall', the Islands of the Foreigners. This last is a reference to their colonisation by the Vikings from the middle of the ninth century onwards. But perhaps most intriguing and perplexing are two Irish terms dating from about the same period. Chroniclers wrote of the Finngaill, the White Vikings, and the Dubgaill, the Black Vikings. Scholars speculate that the references may be to colouring, and that the Finngaill were northern Scandinavians while the Dubgaill were Danes. Others are less literal and interpret the names as signifying old and new. The ships of the White Vikings were seen off Irish coasts first, presumably in the early 800s, and they were followed by the Danes, the Black Vikings. Given that this marches in step with such written records as exist, this reading seems more likely.

The word 'Viking' came into modern English usage relatively late, not until the end of the eighteenth century, and is itself the subject of continuing controversy and confusion, but the consensus appears to be that the Old Norse term 'Viken' was used for the region around modern Oslo, and meant something like 'the Men of the Fjords'. Which brings the argument back to Lochlannaich. However all that may be, 'Viking' came to mean a sea-raider, dangerous, ruthless and Godless, a predatory pirate whose sleek ships could loom over the horizon at any time. As the raiders rowed hard for the beaches, their oars clunking in the rowlock holes, all who saw them in time fled.

The terror they induced is difficult to play down, and it produced some extraordinary incidents – and sacrifices. After Abbot Cellach had established a new monastery at Kells in Ireland, safely inland and beyond the reaches of the Lochlannaich, Iona appears not to have been entirely abandoned. A remnant community held on and despite the fear of more landings, they tended the sacred precinct and maintained a continuity as the heirs of Columba.

In 824 an aristocratic monk called Blathmac left Ireland to go and live at the old monastery. He became its abbot. A near-contemporary account compiled at Reichenau in Germany makes it clear not only that the monk was aware that the Vikings could return but that he embraced the danger. It seems that Blathmac actively sought martyrdom – and he was most certainly not to be disappointed. Perhaps his piety convinced him that a blood sacrifice in defence of the sanctity of Iona might turn God against the savages. But nothing could have prepared the monk for a sacrifice that would be made to another god.

The Reichenau account tells of a dawn raid, but it apparently did not come as a surprise. Lookouts had given the abbot advance warning and he sent away those monks whose courage wavered 'by a footpath through regions known to them' to hide in the interior of the little island. What followed was a scene of terrible violence. The Vikings broke into the monastery and slaughtered almost everyone they came across, except the abbot. Demanding to know where Iona's treasures were buried, they questioned Blathmac. Probably when no answer was forthcoming, they then performed the terrible rite of blood-eagling. As a sacrifice to Odin, they stripped off the abbot's habit and a warrior used a sword to make deep cuts on either side of Blathmac's spine, hacking at his ribs. The rib-cage was then pulled out to form the blood-eagle. It is to be hoped the victim of this appalling martyrdom died of shock early in this grisly ritual. And it was not unique. In November 869, King Edmund of East Anglia died in the agonies of blood-eagling, and others suffered the same fate in Orkney.

The eighteenth-century use of the word 'Viking' coincided with what is known as the Viking revival. In 1814 Sir Walter Scott visited Shetland and, much taken with the dramatic sweep of the landscape, he decided to set the opening of his novel *The Pirate* there, at a place he called Jarlshof, the site of the ruined house of a seventeenth-century laird. It sounds authentic (Scott had an excellent way with names), but in fact it was invented for the novel and was afterwards taken on as the actual name of the place. Jarlshof is now one of the best known Viking sites in the north. Richard Wagner added a soundtrack. As the themes of his operas swirled around the thunder and lightning of the Nordic gods, myths and sagas, theatre designers came up with the classic image of horned helmets, braided hair and glinting weaponry. The popular image of the Viking as a barbarous but occasionally noble and daring savage was born.

The horned helmets have no basis in history or archaeology, and it is likely that Vikings were in the habit of shaving their heads, apart from a little tuft at the front. Like most warriors, they wore iron pot helmets that protected them and did not sport horns stuck on the sides that might have seriously damaged those fighting alongside them. But the barbarity is not ahistorical. Despite the fact that the Vikings initially left little in the way of written record to speak for themselves, and despite the efforts of modern historians to broaden the picture and place emphasis on the cultural and economic lives of these people, the theme of extreme violence rumbled on in the background. The shock of first contact has never completely faded.

While the images and incidents remain graphic, the reasons for that first contact are much more blurred. Land hunger, population pressure or a compelling combination of the two have been advanced as theories, but little evidence of either seems available. In any case, neither is a motive for the initial period of raiding and returning with loot, and would apply only to the later settlements. Vengeance is one of the more unlikely hypotheses. Certain scholars see the ferocity of the Viking attacks as a reaction against the efforts of Charlemagne and his pious advisers such as Alcuin of York to convert forcibly all pagans to Christianity. In reality, motives are usually much more messy and chancy. It may well be that sea-lords first set sail out of the sight of land to cross the North Sea without any clear purpose. When they set a course in a direction they called westoversea, they may simply have wanted to see what lay over the horizon. Perhaps it was nothing more than curiosity and a sense of adventure that sent the longships out into the North Sea. And at the end of the eighth century, there cannot have been blanket ignorance. Sea-lords probably knew or had heard tales of a long island in the west. And were there also rumours of what they might find there? Far-off lands and treasure or opportunity of some sort are often linked.

When the first raiders returned home, what was their impact in the settlements that clung to the shores of the fjords? It is a natural impulse to focus on the impact of the Vikings on Britain and Ireland, but when the ships that rasped their keels up on the shingle beaches of Lindisfarne came back, how were they received? Warriors will have disembarked carrying undreamed-of wealth, with gold, silver, jewel-encrusted gospels, fine cloth, coins, all manner of portable loot and perhaps even slaves, all glinting like treasure that had been there for the taking. Those who had not yet ventured westoversea will have been much motivated.

'Gold-giver' was one of the most positive epithets that could be attached to the leader of a warband, and successful sea-lords bound the loyalty and energy of their warriors to them by the distribution of plunder. It was part of a way of life, something that remained ingrained in Viking culture. Here is a remarkable passage that tells how Svein Asleifsson, a man of Scandinavian descent whose family had settled in Orkney, probably since the second half of the ninth century, organised his year. This annual cycle took place in the middle of the twelfth century and it shows how hard old habits died:

This is how Svein used to live. Winter he would spend at home on Gairsay, where he entertained some eighty men at his own expense. His drinking hall was so big, there was nothing in Orkney to compare with it. In the spring he had more than enough to occupy him, with a great deal of seed to sow which he saw to carefully himself. Then when that job was done, he would go off plundering in the Hebrides and in Ireland on what he called his 'spring trip', then back home just after midsummer, where he stayed until the cornfields had been reaped and the grain was safely in. After that he would go off raiding again, and never came back till the first month of winter was ended. This he used to call his 'autumn trip'.

In the beginning, before settlement started, sea-lords probably sailed westoversea because they could. Centuries of development had honed extraordinary woodworking skills in the fjords and by the eighth and ninth centuries, Viking shipwrights could build very graceful, versatile and sleek ships capable of tremendous speed. What history remembers are the sailors and sea-lords who stood in their prows, but the inventiveness of those who made these beautiful ships also deserves recognition. And it seems that their skills eventually made the voyage westoversea.

In 2009 Graeme Mackenzie was thinking about potatoes. Fifty metres from his croft on the Rubh' An Dunain, a peninsula at the south-western foot of the Cuillin on the Isle of Skye, lay a likely area of rough grazing. But to make a proper potato patch, it needed draining. Having hired a mechanical digger, the main drain was immediately put to use as the jagged ridge of the Cuillin pierced the clouds blowing in off the Atlantic and rain lashed the coastline. When the weather improved, Graeme went back to his ditch and noticed something out of the ordinary. Sticking out of the peaty earth was a 10 cm iron spike. With a shovel, he levered out the unmistakable shape of an anchor. Forged from iron, it measured more than a metre from top to bottom and a metre across from one curved tip to the other. When archaeologists examined the anchor, it proved to be about 1000 years old. Graeme Mackenzie had found something that had been stowed in a Viking longship.

It turned out not to be a stray find. Aerial surveys of the site revealed the outline of what was probably a Viking shipyard on Skye: a stone-lined canal, a quay and a small inland loch called Loch na h-Airde (the upland loch). Boat timbers were recovered from the anaerobic soil and the layout of the canal, quay and loch suggest not only a focus for maritime construction but also a safe haven for ships that needed maintenance. Although Graeme Mackenzie knew what he had found, he could have had no idea that it would lead to the discovery that his croft lay on an ancient shipyard.

Drekar, or dragon-ships, was how Viking longships were described by those who sailed them and those who had good reason to fear their coming. Fixed to their prows were carvings of animals of menace, mythical and real, dragons and snakes. Not only were they intended to terrify human beings, sea-lords also hoped that they would ward off the monsters that lurked in the deep. However, there were much more sophisticated aspects of the construction of these beautiful ships that made them a real threat, and much skill will have been in evidence at Loch na h-Airde.

Drekar were designed principally for speed and versatility. With a shallow draught, they could safely navigate water only a metre deep. This allowed tremendous reach, access a long way upriver and deep inland to attack unsuspecting churches and communities. And the draught also made attacks on coastal targets more effective, preserving much of the element of surprise. Sea-lords could roar on their oarsmen to row hard and keep momentum as the steersman attempted to ground the ship on a shelving beach. It is thought that drekar could reach a speed of up to ten knots when propelled by oars.

They were also double-ended, with symmetrical bow and stern so that they could reverse quickly without the cumbersome business of going about, and that helped them avoid foundering on undersea rocks, sandbanks or ice floes. And drekar were light. This meant that they could be carried by their crew, or for bigger boats, pushed on log-rollers over an isthmus to save a longer journey around by sea. The placename of Tarbert in Kintyre (and elsewhere in the west of Scotland) means an 'overbringing', a place where a portage could be made.

Such virtuoso seamanship was made possible by the skills of Viking shipwrights and their precise understanding of the properties of wood. Unseasoned oak was much favoured and freshly cut timber will have been brought in quantity to Loch na h-Airde on

Skye. Using an impressive toolkit of different sorts of axes, adzes, draw-knives, chisels and twist-drill bits, the shipwrights shaped the oakwood in particular ways. If possible, a single tall tree was preferred for the keel and the overlapping strakes or planking for the hull was always cut with the grain of the wood. This imparted great strength and flexibility and when battered by heavy seas or running against powerful tides, the shell of a drekar could twist a long way out of true without breaking up. The tapered wooden pins known as treenails and iron rivets were used to sew the strakes together and the internal frame stiffened the whole structure. It is thought that oarsmen rowed while sitting on their sea-chests, and to prevent the waves from splashing over the low freeboard, they were in the habit of arranging their round shields along the sides of the ship. That visual cliché is at least accurate.

Sail came late to the north. It seems that masts were only fitted in longships from the seventh century onwards, but they did extend their range greatly. The use of sail enabled the Vikings to make longer voyages and eventually to cross the North Sea. Drekar were open-decked, and if they were to be at sea for some time the men wore primitive oilskins to keep as dry as possible. These were fleeces turned inwards, with the skin side smeared with fish oil to make it waterproof. If there was no wind and the voyage was being made in summer, the stench must have been strong. But the open deck made the drekar vulnerable to swamping. A huge wave could fill a longship in an instant, and in a gale in the Mediterranean in 1992, a replica was capsized and sank. None of that crew were lost, but if such a thing had happened in 792 in the middle of the North Sea, all aboard would probably have been drowned.

From the ninth to the twelfth centuries, Europe's climate appears to have been unusually benign. On clear days and nights, sea-lords will have navigated by looking upwards at the positions of the sun and the stars. Even when it was cloudy, Vikings are believed to have used a sunstone, a type of feldspar capable of polarising sunlight and fixing its source. They also understood the run of the sea, the significance of different wave patterns. In Norse mythology, the nine daughters of Aegir and Ran, a giant and a goddess, were named for the various characteristics of waves observed by sea-lords. Pitching waves, billows, risers, combers, waves with foam-flecks on their crests, transparent waves, red-brown sea-foam, frothing waves, welling waves and cool waves were all commemorated in the daughters of Aegir and Ran.

Horizon Boards

Historians believe that the Vikings used a primitive astrolabe to plot their courses when sailing out of the sight of land. Used by the 'kendtmand', the man who knows the way, a board somewhat like a sundial was used in conjunction with a piece of spar or a sunstone. The latter could change colour to allow the sun's position to be worked out and in the summer in the north, when there was little or no darkness and the sun did not slip below the horizon, this method of navigation could have worked well.

Natural phenomena also helped them plot the right course. What are known as flyways, the habitual flightpaths of migrating birds, and the movement of sea creatures such as whales, shoals of fish and seals formed part of a well understood body of Viking sea-lore. Its signs and patterns were approximate and variable, but navigation was not a precise science and for those sailing westoversea the 800-mile-long target of Britain will have been hard to miss.

And many did hit it. Shetland and Orkney were nearest and it appears that the Northern Isles were the first to be settled by Viking colonists. In contrast to the steep-sided fjords of western Norway and the narrow strips of cultivable land along their shores, the green fields and wide pastures of the two archipelagos will have seemed like a prize. Shetlandic soil is poorer and the northern districts of the Shetland Mainland and the northern islands of Yell, Fetlar and Unst were more suited to pastoralism whereas Orkney, by contrast, could grow more grain and other arable crops. Its position at the edge of the continental shelf compensated because it made the waters around Shetland more nutrient-rich and the fishing grounds consequently more productive.

After the first few summers of raiding, it is likely that Viking sea-lords began to establish bases on Orkney and Shetland as they sailed further south and west in search of easy targets. These ship-camps or longphorts almost certainly grew into permanent settlements as the early groups of warriors increasingly opted to over-winter. Perhaps some men took native women as wives, a comfort of various sorts as the winds whistled in off the ocean.

There must have been resistance. The successors of the king of Orkney who submitted to Claudius in Colchester appear to have been powerful. The broch-village at Gurness was not an unsophisticated cluster of huts around a stronghold but a well-built complex that may have been a royal residence. And the lords who employed the broch-builders at places like Mousa on Shetland certainly commanded significant resources. The Vikings knew these people as Picts. Their name for the Pentland Firth, Pettlandsfjodr, means the Fjord of the Picts. But it seems that the original populations of both archipelagos have been almost entirely effaced. At the outset of the Viking occupation, the native peoples will probably have been worked as slaves or bondsmen of some kind, tied to work on land that their kindreds originally possessed. As the colonists became established and began seriously to develop the business of slavery, more Orcadians and Shetlanders may simply have been removed. It is now very difficult for geneticists to trace native prehistoric DNA. With Viking DNA, it is a completely different story.

In Orkney, 20% of the modern male population carry the classic Viking Y-chromosome marker of M17, and amongst those with older surnames such as Clouston, Rendall, Isbister or Flett, the proportion climbs to 35%. The ancestors of the majority of men living in Orkney now arrived after the Viking settlement, especially following the absorption of the Northern Isles into the kingdom of Scotland in 1468, and consequently they carry markers that came from the south.

Svein Asleifsson, the sea-lord who routinely raided in spring and autumn trips, was described by the great Orcadian author Eric Linklater as the Ultimate Viking. Although he nominally owed allegiance to the twelfth-century earls of Orkney, Svein appears to have acted completely independently. From his island fastness of Gairsay (and its drinking hall large enough for 80 men), which stands sentinel at the mouth of Wide Firth, the bay at the heart of the archipelago, he set sail for plunder. Ranging far down the western costs of Britain, the Gairsay longships attacked the coast of Wales, laid siege to the island of Lundy in the Bristol Channel and took Svein to the Isle of Man to seek a bride so that he could seal a strategic alliance with a Manx sea-lord. Eric Linklater was only echoing the Orkneyinga Saga of *c* 1200, for its composer hailed the Ultimate Viking as 'the greatest man in the Western Lands, either in olden times or present day'.

DNA research suggests that when the last real Viking died in an attack on Dublin in 1171, his marker carried on – with great vigour. A new sub-type of S21, S375, is now prominent in the North Isles of Orkney, on the five major islands north of Gairsay. It is also carried by 30% of men with the surname of Gunn who have taken DNA tests. Tradition, genealogy and history all begin to come together to form a narrative.

It may be that the prevalence of S375 on the islands of Rousay, Westray, Eday, Sanday and Stronsay is linked to the well-attested phenomenon of social selection, where powerful men in the past sired many children with different women. In that way their Y-chromosome markers were spread widely and quickly, much faster than if they had remained monogamous. And few men were more powerful in twelfth-century Orkney than Svein Asleifsson.

The Ultimate Viking's second great fortress was in Caithness at a place the sagas called Lambaborg, thought to be modern Freswick, on the coast to the south of John o' Groats. Tradition holds that Svein's son, Andres, sired a boy known as Gunni and that he was the name-father and founder of Clan Gunn. At one point, their ancestral lands included all of Caithness. The inheritance of Gunni's son, the wonderfully named Snaekoll (it literally means Snow-Head, more colloquially, white-haired) was swelled by lands from his mother, Ragnhild, the daughter of Harald Maddadsson, the earl of Orkney.

Now, the pieces begin to fall into place if 30% of the men of Clan Gunn are indeed the descendants of their name-father, and ultimately of the Ultimate Viking, Svein Asleifsson. Given the prevalence of S375 in the islands around Gairsay, it seems likely. And there is a persuasive comparison. In Clan Donald the lineage of its founder, Somerled the first Lord of the Isles, appears to have come down the generations uninterrupted, since all the MacDonald chiefs share the same DNA marker. And more than 20,000 men around the world with the surname of MacDonald or its variants also carry Somerled's marker. Another example of the effects of social selection.

After *c* 800, Vikings sailed east through the Baltic Sea as well as west, their longships penetrating deep into the heartland of what is now Russia, through the great river systems of the Dnieper, the Dniester and the Volga. But two recently discovered sub-types in M17 turn out to be much more closely linked with Norway. In the modern population, around 12% of men carry S200 and about 8% have S223. Both are also seen in Scotland, brought here across the North Sea in the drekar.

What is surprising is their distribution in Norway, and this offers a hint of definition, of more precisely where some of the Vikings who settled the lands westoversea came from. S223 is not common across the whole of Britain at just under 1%, but it is heavily focused in Shetland at about 10% of men in the modern population, 8% in Orkney and around 3% in the Western Isles and in the north-west mainland of Scotland. In Norway, this marker reaches a peak of 18% in two districts directly north of Oslo, Hedmark and Oppland, and it is not common on the west coast at 3%. This is surprising, since this area with its long North Sea coastline deeply indented with fjords and their sea-going communities lies directly opposite the Northern Isles.

The brother-group of S200 is also found at 10% of men in the modern population of Shetland and 8% in Orkney, but its frequency in the Western Isles and the north-west coast is higher, at 6%. In Norway, S200 is found in Buskerud, Telemark and Rogaland, regions in the south, as well as in the north in More and Oppland. There the frequency is high, at 18% to 22%. On the basis of these comparative figures it is not possible to make secure connections, but it does point to the strong likelihood that those who carry S200 in Scotland are descended from Norwegian Vikings. That marker is found at less than 1% in Denmark and only 3% in Sweden. S223 is less rare at around 8% in Sweden and Denmark and so some of who carry it in Scotland may have ancestors who originated there. But as research continues, it may be possible in the near future to make even clearer links that reach across the North Sea and back across 1000 years.

A fascinating genetic footnote to all of this traffic from east to westoversea is the recent discovery of a deposit of the notorious trade of slaving, evidence of movement in the opposite direction. The British Y-chromosome marker of R1b-S145 is found at a significant frequency in Norway of 5% of men, but it is particularly concentrated along the western seaboard. In the regions of Nordland, Trondelag, More and Romsdal, and Hordaland, R1b-S145 is seen in between 12% and 19% of the modern population. And the bias is firmly Scottish and Irish, with samples clearly coming from a Pictish sub-group, from the Irish group of R1b-M222 and from another Scottish/Irish sub-group. This genetic evidence is underpinned by documentary reports of sea-lords returning to Norway with human cargo on board. As often, absence can be eloquent. R1b-S145 is significantly less common in Sweden and Denmark, and in addition none of the samples found in Norway are from the recently discovered eastern European branch of the marker. It seems that a large proportion of men living in the west of Norway are the descendants of Scottish and Irish slaves.

Far over the western horizon, beyond the sunset, history was shifting in the homelands of the displaced. Their captors and their captors' fellow raiders were engaged in making nations, and not always their own. Pressure exerted by Viking raids down the Atlantic seaboard pushed the focus of Dalriadan power eastwards into the centre of Scotland. In 839 at Strathmore in Angus, a Pictish army was cut to pieces by a Viking host and many of the native nobility were killed. It seems that a power vacuum allowed Kenneth MacAlpin famously to establish himself as a king of Scots and Picts in 843, and his dynasty to flourish thereafter. The realm known as Alba to Gaelic-speakers and later as Scotland was forming, forged by Viking steel.

Recent DNA analysis strongly suggests that the slaughter at Strathmore did not destroy a Pictish identity, it submerged it. A new Y-chromosome marker has been discovered and shown to have arisen amongst the direct prehistoric ancestors of the Picts. In a sample of 3,000 British and Irish men, geneticists have identified R1b-S530 and found that it was 10 times more common in men with Scottish grandfathers than it was in men with English grandfathers. Of 1000 Scottish men tested, 10% carried R1b-S530 while only 0.8% of Englishmen had it. This difference is highly statistically significant and can be applied to the general population. It is clear evidence of a very Scottish marker – but also evidence of much more.

The pattern in Ireland is eloquent. About 3% of Ulstermen carry R1b-S530 but it was only seen once in a sample of 200 men in the rest of Ireland. It seems very likely that its presence in Ulster is the genetic legacy of the plantation of Lowland Scots in the sixteenth and seventeenth centuries. This distribution is often seen with markers that appear to be restricted to Scotland.

Ancient Pictland is usually defined by historians as the area where the enigmatic symbol stones and Pictish placenames such as those that have the prefix *Pit* or *Pett* are found. This heartland lies in Scotland north of the Forth and stones and the *Pit*-names are seen across much of Fife, Perthshire, Tayside, the north-east and around the Moray coastlands.

Crucially, it seems that within Scotland there is a strong concentration of the R1b-S530 marker in those same areas. In central Scotland its frequency is 17% and in north-east Scotland it is 14%. It declines towards northern and western Scotland (8%) and southeast and south-west Scotland (6%). The difference between central and Grampian regions and the rest of Scotland is again statistically significant – and this marker is yet to be detected outside the British Isles as a whole.

The meaning and origins of R1b-S530 unpack a fascinating story. A marker that is very common and widespread in Scotland implies that it has been there for a long time, and R1b-S530 is estimated to have arisen about 3000 years ago. This strongly suggests that it was common amongst the ancestors of the Picts, some of the original inhabitants of Scotland. And it seems that their descendants are living there still. Some Scots, or at least some Scottish men, have not wandered far over the last few thousand years. If they had, then this lineage would be much more common in England. The Vikings may have crushed a Pictish army at Strathmore and destroyed its leadership, but the ordinary lineages of ancient Pictland appear to have lasted.

By contrast, in Ireland the influence of the Scandinavians was limited. Because Ireland's power structures were comparatively diffuse and localised, there were no great dynasties to overthrow and no existing wide-ranging administrations to hijack. Conquest had to be piecemeal. Having established their fortified coastal bases still recognised in placenames such as Wexford, Wicklow and Dublin, the raiders and traders made little lasting impact beyond their wooden stockades. However, where they did settle, they flourished. With one brief interruption, the Norsemen ruled in Dublin for three centuries. They called themselves Ostmen or Eastmen, and after the Anglo-Normans seized Dublin in 1171, they migrated to the north bank of the Liffey where their name is commemorated in the placename of Oxmantown, which derives from Ostmantown. They also left other traces, for near Dublin Castle archaeologists have discovered a characteristic site, a thingmote or a raised mound. Around 12 metres high and 73 metres round, it was where the Viking sea-lords assembled to hear legal judgements and agree new laws.

At the mouth of the River Liffey, close to where Dublin Castle now stands, one of the largest and most lucrative slave markets in western Europe was established. Known in Norse as Dyflin, from the original Gaelic Duiblinn (which translates as Blackpool), the market and its surrounding settlement was the capital place of Olaf the White, King of Dublin.

As late as the eleventh century, the slave trade continued at Dublin, and here is an extract dated 1066 from the 'Life of St Wulfstan':

> There is a maritime town called Bristol, which is on the direct route to Ireland, and so suitable for trade with that barbarian land. The inhabitants of this place with other Englishmen often sail to Ireland for the sake of trade. Wulfstan banished from among them a very old custom which had so hardened their hearts that neither the love of God nor the love of King William could efface it. For men whom they had purchased from all over England they carried off to Ireland; but first they got the women with child and sent them pregnant to market. You would have seen queues of the wretches of both sexes shackled together and you would have pitied them: those who were beautiful and those who were in the flower of youth were daily prostituted and sold amidst much wailing to the barbarians.

Not Viking

In 2011 researchers heard a faint echo of the dismal trade carried on in Dublin. The popular Scottish broadcaster and comedian, Fred Macaulay, took a DNA test confident that it would show up his Viking ancestry. Existing databases had already recorded the results of other men with the same surname and they had indeed been carriers of Norse genes. From the Outer Hebrides, the Macaulays are traditionally seen as one of the Norse clans and the Gaelic rendition of Fred's name is MacAulaidh, or Mac Olaf, the son of Olaf. But when his results were analysed, it was discovered that he had a southern Irish marker. R1b Eoganacht is not only from Munster, it is also the genetic signal of the royal family who ruled the province in the ninth and tenth centuries. The Eoganacht morphed into the princes of Desmond and lasted until the beginning of the seventeenth century.

By far the most likely set of historical circumstances behind this surprising result link to the centuries of the Viking slave trade. Sea-lords often raided across the rest of Ireland for captives, and a man from a royal house would have been likely to fetch a high price at the Dublin market. Once he was put up for sale, Fred's ancestor was almost certainly bought by a Viking lord from the Hebrides who was probably called Olaf or was from the household of Olaf. He then sailed back north with his new and prestigious acquisition. At some point the Irish slave had sex with a Macaulay woman and his DNA marker, R1b Eoganacht, was insinuated into their lineage in that way. It is the most plausible explanation and it appears to fit the facts. Fred Macaulay was delighted.

In 870 Olaf raided and conquered the Old Welsh-speaking kingdom of Strathclyde, undoubtedly selling off its people in his slave market. In the aftermath of the fall of Strathclyde and the subsequent extension of his political reach to Scotland, Olaf and one of his leading allies quarrelled. Ketil Flatnose was almost certainly the Viking ruler of the Hebrides in 870. Viking possession of the islands is remembered in the Gaelic name for Lewis, Harris, the Uists, Benbecula and Barra, for they are still described as the Innse

Gall, the Islands of the Foreigners. And almost all the placenames of the Hebrides, the likes of Stornoway, Tarbet (or Tarbert) and Uig, are Norse in origin.

When the chronicler sat down to write of Olaf's expedition to Strathclyde in the monastic Annals of Ulster, he noted that the king led a fleet from Dublin that also included the ships of 'the Foreigners of Alba', meaning the Hebridean Vikings of Ketil Flatnose. These warriors were much feared. They were also called 'Gall-Gaidhil', the Scandinavian Gaels, and appear to have been the children of mixed marriages between Vikings and native women. By the middle of the ninth century, bands of these warriors were raiding out of the Hebrides (and probably Galloway – named after the Gall-Gaidhil) and attacking communities as far south as Munster.

DNA testing has detected this distinctive group. Their marker is R1b-S68 and at the time of writing they make up 0.5% of all modern Scottish men, approximately 12,500 descendants of the feared Gall-Gaidhil, the sea-warriors who terrorised the Atlantic coastlands.

It seems likely that in 871 Ketil Flatnose and Olaf the White disagreed in Dublin over the lordship of the Innse Gall. It quickly escalated into a family squabble and Olaf's queen became involved. The wonderfully named Aud the Deep-Minded was also Ketil's daughter, and when the dispute stalled into a stand-off and the Hebridean Vikings rowed out of the Liffey to return home, Olaf's queen went with them. It was a divorce that would have, in every sense, far-reaching consequences.

On her voyage home to her native Hebrides, Aud the Deep-Minded must have considered her future, and the options for a divorced queen of Dublin cannot have seemed numerous or even appealing. Ketil Flatnose's grip on his Hebridean realm weakened after 871 and it seems that he died not long after his confrontation with King Olaf. Aud's son, Thorstein the Red, began to assert himself in the far north of mainland Scotland. According to the Laxdoela Saga, he carved out a lordship in Caithness and over much of Sutherland (the land to the south of the long-standing Viking possessions of Shetland, Orkney and Caithness). But in the decade after 880 Thorstein overreached himself and was killed by native Scottish warlords. Having left her royal husband and lost their son,

Aud the Deep-Minded responded with the instincts of her Viking ancestors. She commanded that a ship be built and planned a voyage.

Aud's expedition was ultimately to result in new colonies on Iceland, which later led to relatively shortlived settlements on Greenland, and eventually the Viking discovery of North America. Her sea-path to Iceland followed traces left by northern hermit-monks who had rowed and sailed their curraghs to leave their name on islands in the Hebrides and beyond. Known in early Gaelic as 'papar', which is cognate to pope and meant a father or a priest, these monks made landfall and set up windy shelters and cells at places whose names now recall them: Pabbay and Bayble in the Western Isles, Papa Westray in Orkney and Papa Stour in Shetland. And off the southern coast of Iceland lies Papey, a small island where monks lived out austere lives. Placenames also detail a secular Irish presence in the North Atlantic. 'Vestmen' is the Old Norse word for Irish Vikings (as opposed to Ostmen for Scandinavian Vikings or 'Irar', the name they gave to natives and slaves) and at Vestmanna on the Faroe Islands, they left a mark on the map.

In 902 the Irish managed to expel the Vikings for a time. Consequently there is no Scandinavian DNA legacy to speak of since settlement was so sparse and temporary. And in Wales impact was also slight. Only a scatter of placenames such as Swansea (Svein's Sea) and the islands of Skokholm and Skomer remember the raids of sea-lords like Svein Asleifsson.

It was different in England. The entry in the *Anglo-Saxon Chronicle* for the year 865 noted the arrival of what it described as the Great Heathen Army, as it drove up the keels of its fleet onto the beaches of East Anglia. Led by princes, the sons of Ragnar Lodbrok (who may have been a Swedish king), the army had not been mustered in Norway. These warriors were probably the Black Vikings of the Irish annals, men from Sweden and Denmark. Their intention was not to seek temporary plunder, to fill their ships with loot and depart. The Great Heathen Army had come to England to conquer.

No firm numbers or even estimates were recorded but the army of the Black Vikings was large enough to defeat the forces of several Anglo-Saxon kings. Usually referred to as Danes, these warriors quickly exchanged their longships for ponies. In order

The Skraelings

Ten thousand years after the end of the last ice age, Viking sea-lords came face to face with a people whose culture closely resembled that of their own ancient ancestors, the pioneers who repopulated Europe. It was a moment when history came full circle. Sixty thousand years after they crossed the Red Sea, our species had finally girdled the Earth. Those whose ancestors had turned west to populate Europe met the descendants of the men and women who had kept moving eastwards, crossed the great landbridge of Beringia and peopled the Americas. The Vikings named them 'the Skraelings', meaning the Skin People, or more precisely, the people who wore skins.

After the establishment of the Greenland colony *c* 985 in the warming climate of the North Atlantic, Leif Eriksson fitted out an expedition and set a course westward. He and his sea-lords had heard tell of a distant coastline where tall trees grew, and wood was a scarce resource in windy Greenland. Others followed and briefly settled in the land that became known as Vinland the Good.

Perhaps the best documented expedition was led by Thorfinn Karlsefni, and the sagas had much to relate. It seems that Greenland had been virgin territory when the Vikings arrived, but Vinland the Good was not. Here is a passage from the Saga of the Greenlanders:

> After the first winter, summer came and they became aware of the Skraelings, who came out of the forest in a large flock. They carried packs which contained furs and sables and pelts of all kinds ... Then the Skraelings put down their packs and opened them up and offered their contents, preferably in exchange for weapons; but Karlsefni forbade his men to sell arms. Then he hit on the idea of telling the women to carry milk out to the Skraelings, and when the Skraelings saw the milk they wanted to buy nothing else.

This a fascinating account of the meeting of two very different cultures, a moment when the prehistoric past emerged from the forest to interact with the medieval future.

to give themselves the ability to surprise their targets and move as quickly over land as they could over water, the leaders of the Great Heathen Army negotiated a truce with King Edmund of East Anglia. It involved him handing over many ponies so that oarsmen could become cavalrymen. Or at least dragoons, men who would ride to battle and dismount to fight. Led by three brothers, Ivar the Boneless, Halfdan Ragnarsson and Ubbe Ragnarsson, the Danes consolidated at first, over-wintering at Thetford in Norfolk. While warriors learned how to ride and break in whatever mounts Edmund's men had supplied, their warlords gathered intelligence and laid their plans for the campaigning season to come.

In 866 they rode north. Having learned that it was a kingdom riven and weakened by civil war, they attacked Northumbria in the late summer, and in November they took the old Roman city of York. More than three and a half centuries after the remnants of the Sixth Legion had marched through the gates for the last time, long after the rule of the Duke of the Britains, York's imperial past was not yet entirely fled. At the confluence of the navigable rivers Ouse and Foss, the old fortress dominated the landscape, its walls and buildings the sole stone structures to be seen. At the principia, the headquarters of the legionary fortress, there stood an impressive basilica. Its colonnaded hall was altered in the fifth century and there is evidence that it stood for some time afterwards. Under York Minster, on the same site, great columns were found toppled and broken. No record of the reaction of the warriors of the Great Heathen Army has survived but this ninth-century Old English poem offers a sense of wonder at what they may have seen in York in 866:

The Ruin

Splendid this rampart is, though fate destroyed it,
The city buildings fell apart, the works
Of giants crumble. Tumbled are the towers,
Ruined are the roofs, and broken the barred gate,
Frost in the plaster, all the ceilings gape,
Torn and collapsed and eaten up by age.

There is evidence that the Danes tumbled some of the towers built along the run of the Roman walls. Perhaps they were robbing the dressed stone to plug gaps in the defences. Ivar, Halfdan and Ubbe seem to have established their headquarters at the western gatehouse of the fortress. Recorded as late as the fourteenth century, the Old Norse placename of Konungsgurtha was used for what is now King's Square, an English translation and an area hard by the old Roman gate into the city. The rectilinear street layout appears still to have been used but as buildings decayed, it was eventually criss-crossed and is now difficult to recognise.

After several effective summer campaigns against Northumbria, East Anglia, Mercia and Wessex, the Great Heathen Army had overcome all opposition except the forces of Alfred the Great. Only Wessex, the embryonic kingdom of England, survived. In 874 the Danes divided their army. Halfdan led his men north and by 876, chroniclers reported that he shared out land amongst them and they 'ploughed the land and supported themselves'. Meanwhile Guthrum, another Viking warlord, led the southern Danes to the conquest of much of eastern England until his advance was halted by Alfred at the Battle of Edington in 878. The resulting Treaty of Wedmore formally partitioned England south of the Humber between Wessex and what became known as the Danelaw, the region where the laws of the Danes were paramount.

In political terms, this Scandinavian occupation of more than a third of the landmass of modern England lasted for only 70 years, except in the area now covered by Yorkshire. In 954 the last Viking king of York, Eric Bloodaxe, was driven out and English overlordship reestablished. But despite what appears to have been a temporary hegemony, there is what seems to be a striking genetic legacy from this period, particularly in Yorkshire. It appears that Yorkshire and the north of England have a significantly higher proportion of the Scandinavian marker S142 than any other region of Britain or Ireland.

From the end of the Roman Empire in the west and the arrival in Britannia of Anglo-Saxon warbands, the composition of incoming settlers changed. 'Small groups of men in small boats' is a favoured description by more than one historian. The Angles carved out the kingdom of Deira in Yorkshire, and one of its capital places was the old Roman city of York. From the ninth to the eleventh centuries the county was firmly part of the Danelaw, and even after the expulsion of Eric Bloodaxe its cultural character continued to reflect that of the incomers from the east. Scandinavian and Germanic placenames pattern Yorkshire in greater density than anywhere else in England. It is not surprising therefore to see a much higher frequency of Germanic male lineages – 52% of men carry one of this group of markers as opposed to 28% over the rest of Britain.

This striking finding contrasts with the picture of mtDNA in Yorkshire. More than 62% of all female lineages so far tested appear to have arrived soon after the ice melted *c* 9600 BC, and they kept on coming throughout prehistory. Equally eye-catching is the fact that these motherlines appear to have arisen in the south, in Iberia and the western refuges, the painted caves in the steep-sided valleys on either side of the Pyrenees.

What this means is something straightforward. History can indeed leave a clear deposit on our DNA. Many warriors and male settlers came to live in Yorkshire and in other parts of eastern Britain after the fall of the Empire, and while some women did accompany them in their ships, these men appear often to have taken native women as their partners, the mothers of succeeding generations. In Yorkshire this enduring genetic legacy surely allows a little exaggeration. Perhaps the people of England's largest county really can claim to be a different breed.

A forgotten frontier once ran along the course of the River Eamont, a mile or so south of Penrith. A cluster of henge monuments are a reminder of a meeting place used for millennia, and in much more recent times the boundary between Cumberland and Westmorland was fixed on the little river that runs east out of Ullswater to join the Eden. On 12 July 927, a meeting to determine the political shape of Britain was held on its banks, for the Eamont was the northern limit of the realm of Aethelstan, the great and powerful king of England. Grandson of Alfred, he had aggressively expanded out of Wessex and overrun the Danelaw as far as York, later seizing the old Viking city and the lands around it.

The *Anglo-Saxon Chronicle* records that Owain, King of Strathclyde, had ridden to Eamont, the southern bounds of his own much expanded realm. Constantine, King of Alba, of Scots and Picts, also came to the conference and they were joined by Hywel Dda, King of Deheubarth in west Wales, and later of Gwynedd in the north. And they met a lesser man, but someone who was neverthe-less a key figure. Aldred, the last to call himself King of Bernicia and Deira, ceased to reign in 927 and in his stead came Ealdred, the son of Eadulf. He was later to style himself earl of Bamburgh. Aethelstan had driven out Guthfrith, king of York and installed Ealdred to rule on his behalf. Three hundred and seventy years after its founding ruler Ida had begun to reign, the glittering kingdom of Northumbria had passed into history.

A northern chronicler was in no doubt about the purpose of the conference on the Eamont. In return for an agreement on boundaries, the northern and Welsh kings consented to acknowl-edge Aethelstan as Bretwalda, Britain-Ruler. Or in a more flowery version as 'King of the English elevated by the right hand of the Almighty to the throne of Britain'. On coins and inscriptions the old province of Britannia was brought back to life. But whatever the flourishes, the rhetoric and the borrowed glories of Rome, the Celtic kings said one thing to Aethelstan and did another.

In 934 the king of the English was forced to march his army north into Scotland, and three years later he faced a formidable alli-ance. Olaf Guthrifsson, king of Dublin, sailed his warfleet across the Irish Sea to join the hosts raised by Constantine, Owain, Aralt (the king of Man) and perhaps Hywel Dda. At a place called by the chroniclers Brunanburh, probably Bromborough on the Wirral, they attacked a smaller force led by Aethelstan and his captains. Hardened by wars of expansion, the Saxons triumphed and the *Anglo-Saxon Chronicle* abandons its usual terseness:

> . . . the field streamed with warriors' blood
> when rose at morning tide the glorious star
> the sun, God's shining candle, until sank
> the noble creature to its setting.
> As fled the Scots, weary and sick of war
> Forth followed the West Saxons, in warbands
> Tracking the hostile folk the livelong day.
> . . . There lay five kings
> Whom on the battlefield swords put to sleep,
> And they were young, and seven of Olaf's earls
> With Scots and mariners, an untold host . . .

It may have been the last moment when Celtic Britain could have reasserted itself. The anthemic Welsh poem, the 'Armes Prydein Fawr', the Prophecy of Great Britain, called on all of the British to unite with the Vikings to drive the Sais, the hated Saxons, back into the sea that had brought them. But they were too many.

And their pretensions were powerful, persuasive. All sensible conquerors assert their hard-won rights quickly and arrogate to themselves titles that appear to justify them and magnify their importance. After Brunanburh, Aethelstan called himself not only Bretwalda in imitation of Rome but also took the analogy even further when he adopted the rank of Basileus, or Emperor of New Rome. Reaching east across the length of Europe, the king of England borrowed from the Greek-speaking successors of Augustus who sat in imperial state in Constantinople. Aethelstan could think no bigger than that. Elaborate show was an integral element of rule from the glittering city on the Bosphorus, and it also mattered to lesser kings. In an age before mass communication, Aethelstan's heir, Edgar, showed he had a talent for symbolic ritual. In 973, he had himself rowed down the River Dee by no fewer than eight British kings, and perhaps the royal barge glided below the walls of the old legionary fortress at Chester. It was of course an elaborate show of submission, a clear acknowledgement of the political reach of the emperor of the English.

Another new name was coming into currency. 'Engla-land' began to be used by the Wessex dynasts to describe their imperium, but it seems an unlikely choice. An alternative rendering would be Angle-land, so why did the Saxon kings of Wessex adopt it? The British of the west, the Old Welsh speakers and those who described the world in early forms of Gaelic, had no doubts; they talked of Saesneg and Sasunn, the land of the Saxons. Perhaps it was intended as an inclusive term, one that would compass the former lands of the Angles in the east of England, territory that had been settled by many Danes and needed its old identity reasserted. But that feels like the application of modern judgements across an interval of 12 centuries. By the end of the tenth century the name of Engla-land was becoming widely accepted, and by the early sixteenth its modern spelling of England was the norm.

Whatever its genesis, the name appears to have had a unifying effect, converting the kings of Wessex into something grander and more enduring. Nevertheless, the propaganda did not blind Edgar's legislators to the ethnic realities, for law codes made careful distinctions about 'all the nation, whether Englishmen, Danes or Britons'. As late as the twelfth century in Scotland similar respect for cultural and linguistic differences was enshrined in 'leges inter Brettos et Scotos', clearly still seen as separate communities who spoke Old Welsh and Lowland Scots. Particular legal concepts were noted down in recognisable Celtic terms.

If England was made in the tenth century by the House of Wessex, it was very nearly unmade or at least absorbed into a greater imperial entity in the eleventh. In what historians call a Second Viking Age, raids restarted with a vengeance. In 980 longships sailed up the Dee in the wake of Edgar's ceremonial barge and attacked Chester before raiding inland, causing tremendous destruction. Fleets arrived in 1002 and Aethelred the Unready found himself fighting the wars of his ancestors once more. So intense was the pressure that he was humiliated into paying Danegeld, and as the keels of the raiders came onshore, frustration boiled over. Orders went out from the royal administration at Winchester to massacre all the Danes living in England. Futile from the outset, and impossible where Scandinavian settlement was at its most dense, this desperate thrash inflamed the large population of the old Danelaw into insurrection. It was reported that a princess had been one of the victims of what history came to call the St Brice's Day Massacre. She was Gunhilde, the sister of Svein Forkbeard. Messages sailed east across the North Sea.

In the spring of 1003, Svein I, King of Denmark, stood in the prow of the royal longship as it was rowed into the Humber Estuary at the head of a huge fleet to avenge his sister's death. Another Great Heathen Army made landfall. Aethelred was not Unready in the modern sense, the epithet is Old English for something like 'poorly advised', but he was weak. By 1013, he had been driven into exile across the English Channel to Normandy, forced to give up the new kingdom of Engla-land to Svein's warbands. An Empire of the North Sea came into being.

It promised much, and enjoyed dynamic and practical leadership, but was ultimately destroyed by dynastic failure. Svein died in 1014, and the crews of the longships in the River Trent immediately swore loyalty to his son, Cnut. He also had the support of the men of Lindsey and after a false start, the man who became known to generations of schoolchildren as Canute, established himself as King of Engla-land. Having defeated Aethelred's son, Edmund Ironside, at Assandun in 1016, he did what statecraft dictated and married Emma, Aethelred's widow, converting to Christianity at the same time.

Cnut truly was an emperor, ruler of three kingdoms united by the North Sea: Norway, Denmark and England. But his reign was a historical cul-de-sac, so much of a diversion that the dynasty of four Danish kings of England is sometimes ignored. Cnut died in 1035 and his son and successor, Harold, five years later. His son by Aethelred's widow, Harthacnut, lasted only two years as ruler before the son of Edmund Ironside restored the Wessex line. Edward the Confessor had been raised and educated in Normandy, his mother Emma was the daughter of a Norman duke and his father was half-Norman. These were links that would eventually pull the last great invasion of England across the Channel in what is often seen as another Viking takeover at one remove. But was it?

Gongu Hrolf was huge, such a big man that no pony had the strength to carry him. His name means Rolf the Walker, because that was what he had to do. Born sometime in the 860s, he was the nephew of Sigurd, the first earl of Orkney, and a born adventurer. Probably sailing from the fjords in the 880s, Rolf had become a warrior, a member of one of the Viking warbands who seized land along the banks of the Seine in northern France. It was a time of tremendous change, a time for opportunists.

The Empire of the Franks, reforged by Charlemagne into a new version of Rome and consummated with his coronation by the pope on Christmas Day 800, was breaking up. The western portion, Neustria, had Paris as its capital place but the old imperial counties such as Anjou, Toulouse and Flanders were pulling away from the centre. Semi-independent local dynasties were growing and they bolstered their new-found power with the evolution of what would become known as feudalism. In return for land, the leaders of warbands promised a lord military service. And it worked reciprocally. Lords also offered their vassals protection.

These interlocking loyalties assumed a pyramidal structure. The heirs of the old imperial counts became the king's tenants-in-chief and below them were descending ranks of sub-tenants. These men were given control over localities in return for the supply of an agreed number of soldiers, usually knights, to the host of a tenant-in-chief and through him to the king. Other services and skills were sometimes included in these feudal calculations. What they did was to spread the expense of two new elements in medieval warfare: the armoured knight on horseback and the building of castles. Another factor further galvanised the growth of feudalism in northern France, and this was the pressure exerted by the invading Vikings, amongst them Gongu Hrolf.

Charlemagne's great-great-grandson laboured under the soubriquet of Charles the Simple but his dealings with the Vikings on the Lower Seine suggest subtlety as well as simplicity. In 911 the king of France met Gongu Hrolf at a monastery at Saint-Clair-sur-Epte, north-west of Paris. Bringing into play the new structures and attractions of feudalism, King Charles awarded the Viking warlord the grand title of Dux or Duke, a comforting echo of late imperial Roman nomenclature, while at the same time making him a royal vassal and a tenant-in-chief. In return, the king conceded nothing but a recognition of the political reality as he formally granted Hrolf and his followers a vast tract of land west of the River Epte, which was described by the monastic scribes as Normandia, Normandy, the land of the Northmen. To seal the oath of homage, the man the same scribes called Rollo also agreed to abandon his pagan gods, Thor, Freya and Odin, and become a Christian. It was reported that a mass conversion followed when around 5000 Viking warriors accepted baptism. Many took native Frankish women as brides and they settled into the pattern of French society and politics. Rollo's dynasty was to rule Normandy for 250 years, and in the decades that followed a dunking in the River Epte, it appeared that the Vikings became more French than the French.

The new dukes were loyal to the dynasty of Charles the Simple, and when it failed they did not attempt a coup but instead were enthusiastic supporters of the new candidate for kingship, Hugh

Capet. The royal aspirations of the descendants of Rollo would lie elsewhere. The Normans vigorously defended their king's interests when the Burgundians attacked in the later tenth century and as a reward, the duchy of Normandy was greatly extended to the west. The dukes became patrons of the Church, endowing buildings such as the Abbey of Bayeux. But they remained natural adventurers. In the middle of the eleventh century, Count Roger d'Hauteville led a Norman expedition to Sicily in order to conquer the island and southern Italy in the name of the pope. Under their rule, the extraordinary kingdom of Sicily flourished, its culture heavily influenced by its Muslim and Greek subjects. But those who were left behind on the shores of the Channel were no less ambitious.

William the Bastard had become duke of Normandy in 1035, the year that saw the death of Cnut and the gradual slackening of Scandinavian control of England. Edward the Confessor was William's cousin, and during the long exile of the House of Wessex he had lived in Normandy. It may be that during that time he acknowledged the Norman duke's claim to the English throne. And when Earl Harold Godwinson, a powerful relative (by marriage) of Edward's, was shipwrecked on the Norman coast, the price of his release may have been an oath to recognise William's right to succeed Edward.

When the Confessor died in January 1066, having ruled for 24 years without producing an heir in direct line, Harold seized the throne and the balance of history shifted once more and with gathering speed as a great drama began to unfold. Within the space of less than three decades, England would spin out of a Scandinavian political orbit, back to the rule of a native English dynasty and then fall under the control of Norman dukes. None of these destinies were inevitable.

In the summer of 1066, King Harold II began a frantic campaign to keep his throne. He faced no fewer than three armies who attacked him from two directions to claim the kingdom and remove the usurper. On 26 September at Stamford Bridge in Yorkshire, Harald Hardrada, also called Harold the Ruthless, king of Norway, aimed to regain the empire of Cnut. He was supported by warbands commanded by Tostig, the disgruntled brother of the new king of England. In a brilliant and dashing display of battlefield tactics, the English army surprised and divided the forces of Hardrada and Tostig, and destroyed them. The most spirited resistance was said to have come from one man, a massive axe-swinging Dane who held the bridge over the River Derwent until an English spearman waded under him and stabbed upwards through the wooden slats.

Meanwhile intelligence had reached King Harold that William the Bastard had mustered a third invading army and a fleet waited for the wind at the port of St Valéry. The king's captains quickly marshalled their victorious warriors into marching order and they hurried south from Stamford Bridge to meet the new threat from across the Channel. There followed an extraordinary sequence of events, a drama played out over only a few days, a time that would turn Britain's history decisively.

Writing only 50 years after the Battle of Hastings, William, a monk at the abbey at Malmesbury in Wiltshire, described the hours before the armies clashed on the slopes of the ridge known as Senlac Hill.

The courageous leaders mutually prepared for battle, each according to his national custom. The English, as we have heard, passed the night without sleep, in drinking and singing, and in the morning proceeded without delay against the enemy. All on foot, armed with battle-axes, and covering themselves in front by the juncture of their shields, they formed an impenetrable body . . .

Pioneered by the heavily armoured hoplites of classical Greece and perfected by the disciplined ranks of Rome's legionaries, the shield-wall had been a much favoured battlefield tactic for more than a thousand years. On Senlac Hill, Harold's warriors formed up in a curved line, each man locking shields with his neighbour. 'Rim to boss' was the cry from the older men as shields overlapped, and in the Bayeux Tapestry English infantry are clearly shown obeying orders precisely. Once the wall was locked tight, a second rank closed in behind, and then more men pushed in behind them. When shield-walls clashed, a battlefield became a murderous scrimmage as men shoved and hacked at each other. The phrase 'weight of numbers' was apt.

Shield-Wall Now

This ancient military formation is seen regularly in European cities today. Riot police carry plastic shields similar in shape to those used by Roman infantry but they appear markedly less skilled in forming a shield-wall. Perhaps because there is no boss, they do not lock them in the way that Harold's army at Hastings did and gaps often open up. But the plastic shields are more a protection against thrown objects such as bricks or even Molotov cocktails. To ward off such aerial assaults, the police sometimes form a testudo or tortoise defence where shields are held over their heads. These would also often benefit from the expertise of a centurion and the liberal use of his vine stick.

Harold's army, and especially his housecarls or household warriors, were often armed with double-bladed Danish axes. They could be swung over the shield-wall to deliver a stunning blow or used at knee level to undercut, attack an opponent's legs and disable him. And the men in the second rank could also swing at the men pushing at their own front rank. Most men were right-handed, and when they raised an axe or a slashing sword, they immediately opened their right side to a stab or thrust with a spear. That was the moment when a right-hand man's intervention became crucial as he attempted to deflect a blow and protect his comrade. Shield-walls tended to wheel because of the prevalence of right-handedness. Most of the shoving was done by the left arm threaded through the loops and grips of a shield and this tended to slew two lines of fighting infantry in an anti-clockwise direction. When this began to happen, warlords pulled their men back, desperate to avoid the worst calamity of all – the opening of gaps in a shield-wall. Once an enemy broke through and got behind a line, slaughter almost always followed. Right-hand men who protected their comrades were not only acting altruistically, they were ensuring that the wall was not breached.

As a charge collided with a stationary shield-wall, everything depended on the impact. When moving fast in tight groups, an attack was always much more effective than a scatter of warriors who raced ahead of their comrades. And from Hastings to Culloden, the outcome depended on the shield-wall or the ranks of redcoats standing fast. Lines could buckle and then push back, but if a powerful momentum broke through, the battle was quickly decided. Men fought up close, roaring and grunting inches from each other's faces as axes and swords chopped and stabbed. The Norman infantry on Senlac Hill will have smelled the mead on their opponents' breath. And as William of Malmesbury reported, many men needed mead to give them the courage to stand shoulder to shoulder. When men were struck down or slipped and the wall gave a little, that was an immediate signal for warriors to push forward hard, trampling over the fallen wounded who were finished off by the axes and spears of the second rank. What mattered at all times was the ability to anticipate gaps and plug them. Shield-wall warfare was a tangle of bodies, of men screaming in death agonies, of the stink of voided bowels and above all the bitter taste of blood. But it was effective. William the Bastard had ordered his archers to soften up the men on the ridge with volleys of arrows but they were easily caught on the shields of the English wall. And even a charge of armoured knights could not breach it, their warhorses, the well-schooled, snorting stallions known as destriers, shied to one side as they were spurred up the slope to the bristling line of Harold's warriors.

The great battle at Hastings was a long-drawn-out struggle, and it cannot have been continuous. Men grew exhausted, ragingly thirsty and there must have been lulls in the fighting as the lines regrouped and the wounded were pulled back. William of Malmesbury takes up the story:

> . . . the battle . . . was fought with great ardour, neither side giving ground during the greater part of the day.
>
> Observing this, William gave a signal to his troops, that, feigning flight, they should withdraw from the field. By means of this device the solid phalanx of the English opened for the purpose of cutting down the fleeing enemy and thus brought upon itself swift destruction; for the Normans, facing about, attacked them, thus disordered, and compelled them to fly. In this manner, deceived by a stratagem, they met an honourable death in avenging their enemy; nor indeed were they all without their own revenge, for, by frequently making a stand, they slaughtered

their pursuers in heaps. Getting possession of an eminence, they drove back the Normans, who in the heat of pursuit were struggling up the slope, into the valley beneath, where, by hurling their javelins and rolling down stones on them as they stood below, the English easily destroyed them to a man. Besides, by a short passage with which they were acquainted, they avoided a deep ditch and trod underfoot such a multitude of their enemies in that place that the heaps of bodies made the hollow level with the plain. This alternating victory, first of one side and then of the other, continued so long as Harold lived to check the retreat; but when he fell, his brain pierced by an arrow, the flight of the English ceased not until night.

Leofwine and Gyrthe, Harold's brothers, had also been killed when the shield-wall broke ranks and consequently the English army was left leaderless. It seemed that Hastings turned on the flight of a single arrow. But defeat was not total. Other sources confirm that many Normans were ambushed and cut to pieces by retreating English warriors as dusk fell over the battlefield. The invading army was exhausted and unable to pursue the scattering of those who fled north. Ultimately though, William the Bastard did drive home his advantage and become William the Conqueror, and once he had subdued most of the native opposition, he began to assert an astonishing degree of elite ethnic dominance.

Hereward the Wake

The leader of Saxon resistance was based in the Isle of Ely in the maze and marshlands of the Fens and in 1866 Hereward was immortalised in a novel by Charles Kingsley. His soubriquet may mean 'the Watchful'. He attacked the Norman-controlled Peterborough Abbey and in 1070 took part in the anti-Norman uprising centred in the Fens. He had little success and appears to have become an outlaw. But what brought him posthumous fame and even glamour was the subtitle of Charles Kingsley's novel about him. It was 'The Last of the English'. It was the basis of a superb BBC television series starring Alfred Lynch as Hereward, but evidently not one episode of this series has survived.

Twenty years after the battle on Senlac Hill, between 4000 and 5000 estates had been prised out of native ownership and redistributed to Norman lords. Out of approximately 180 tenants-in-chief, that is, men who held land worth more than £100 a year, only two were English, and in the income level below that, only about 100 out of 1400 were English families. Amongst the rich holdings of the Church, Norman control was just as overwhelming, with just one out of sixteen bishops being English. William of Malmesbury wrote a portrait, not altogether flattering, of the conquering king who engineered this radical shift:

He was of just stature, ordinary corpulence, fierce countenance; his forehead was bare of hair; of such great strength of arm that it was often a matter of surprise that no one was able to draw his bow, which he himself could bend when his horse was in full gallop; he was majestic whether sitting or standing, although the protuberance of his belly deformed his royal person; of excellent health so that he was never confined with any dangerous disorder, except at the last; so given to the pleasures of the chase, that as I have before said, ejecting the inhabitants, he let a space of many miles go desolate that, when at liberty from other avocations, he might there pursue his pleasures. His anxiety for money is the only thing on which he can deservedly be blamed. This he sought all opportunities of scraping together, he cared not how; he would say and do some things and indeed almost anything, unbecoming to such great majesty, where the hope of money allured him. I have no excuse whatever to offer, unless it be, as one has said, that of necessity he must fear many, whom many fear.

William the Bastard was crowned on Christmas Day 1066 in Westminster Abbey, and through an intense programme of castle-building he and his supporters had asserted control of most of England and much of Wales by 1070. The cupidity so deplored by William of Malmesbury was perhaps best exemplified by an accounting document startling in its scale, the Domesday Book, a meticulous compilation of records that confirmed exactly how wealthy the new king and his nobles had made themselves. In 1086 agents surveyed a staggering 45,000 landholdings in 14,000 named locations – it was an exercise unparalleled anywhere else in Europe. And it also demonstrated how complete was the Norman takeover.

The Normans kept coming. When Henry I succeeded William Rufus in 1100, he brought with him a man called Alan fitzFlaad, a Breton knight. Many of his countrymen had come with the Conqueror 35 years before and they had formed the left wing of the Norman army as they charged up the slopes of Senlac Hill. Alan fitzFlaad was made feudal lord of Oswestry in Shropshire. His third son, Walter fitzAlan, changed the fortunes of his family when he met David I, King of Scotland, some time in the 1130s. And it was a meeting that would also change Scotland and Britain.

FitzAlan was appointed dapifer, the first Steward of Scotland, by David I, an office that was made hereditary in 1157. The role of steward (or 'stewart') became a surname and in the fourteenth century, after the failure of the macMalcolm dynasty in the late thirteenth, the Stewarts eventually became kings. When Elizabeth I died without a direct heir in 1603, Robert Carey rode north to Edinburgh to be the first to tell her closest male relative, James VI of Scotland, that he had become James I of England and Ireland. After a turbulent dynastic tenure that included a regicide and the establishment of a British republic in all but name, the last Stewart king, James VII and II, fled London and his throne in 1689 when his daugher Mary and her husband William of Orange advanced on the capital after landing at Brixham in what some might describe as the last conquest of Britain.

The Stewarts did not go gently, and their last chance of restoration only flickered and died nearly 60 years later, with the defeat of the clans at Culloden in 1746. But their DNA lives on and thrives in several noble lineages in Britain. Perhaps the most vigorous is the House of Buccleuch. Descended directly from James Scott, the Duke of Monmouth and half-brother of James VII and II, Richard, Duke of Buccleuch, carries the Y-chromosome DNA of Charles II and a long line of kings of England, Ireland and Scotland before him, stretching right back to Walter the Steward of Scotland. For all that time, the Christian name of Walter has remained in the family. Richard Buccleuch's marker (the Stewart marker R1b-S463) is a sub-type of the Scottish/Irish group R1b-S145 and it was almost certainly carried across the Channel by Alan fitzFlaad, the Breton knight. But Brittany is not the place where that sub-type of R1b-S145 is most common, and it very probably did not arise there. Its highest frequency is found in south-west England, in the modern counties of Wiltshire, Dorset, Somerset and Devon. It was this corner of the old province of Britannia that withstood the incursions of the Anglo-Saxons longest, and from where migrants sailed across the Channel in the fifth century to Armorica. So many left that this part of western Gaul was renamed Brittany, or Little Britain. And the scholarly consensus is that the Breton language, closely cognate with Cornish, arrived on Armorican shores at the same time. Therefore it seems very likely that the Royal Stewart marker first arose not in Scotland nor in Brittany but in what is now south-west England, amongst the Romano-British peoples pushed westward by the Saxons in the fifth and sixth centuries. With their inimitable talent for expressing the essence of a story, British newspaper sub-editors neatly turned history on its head when news of the likely origins of Royal Stewart DNA was made public. One headline ran 'Bonnie Prince Charlie Was English'.

Were the Normans really Northmen? Was the conquest of 1066 in effect the last great Viking raid? There is some evidence that it was not. Men pass on surnames as well as their Y-chromosome DNA and the entries in the Domesday Book showed how these began to be adopted. Most of the early examples were territorial, with 'de' prefixed to a placename by an incoming Norman lord. By 1400 most men and many women in Scotland and England had acquired a surname.

A survey of the Y-chromosome markers of 600 bearers of Norman surnames in Scotland is very revealing. Those men called Oliphant, Chisholm, Riddel, Grant, Stewart, Bruce, Bissett, Corbett, Sinclair, Montgomery, Mowat and others had no enrichment for Viking markers. For example, the Viking M17 marker in Orkney is present in 35% of men with old Orcadian surnames such as Clouston, Rendall or Flett. But in the Norman group of 600 men, M17 was found at a frequency of only 3%, the background percentage in ordinary English and Scottish samples. To the degree that this group is representative of Britain as a whole, and it may not be, there is at present little support for the notion that the Normans have much Norse in them. In fact, small samples from Normandy itself agree with this tentative finding because they also show no enrichment of the M17 marker. The footsteps of Gongu Hrolf are very faint.

What these findings suggest is that 1066 saw not so much a Norman Conquest as a French Invasion.

9

The Royal British

�֍

ALL OVER BRITAIN the descendants of royalty and the nobility move anonymously amongst us, many of them entirely unaware of their famous ancestry. Plantagenets wait at bus stops, Tudors push their trolleys to the checkout, Stewarts stand on station platforms, and Hanoverians park their cars in the multi-storey. A substantial number of those who visit the stately homes of Britain and troop through their grand salons, shepherded by ropes and plastic carpet protectors and who gaze up at portraits of gorgeously robed marquises would be amazed to know that they are related, sometimes closely related.

Over the 1000 or so years since the Norman Conquest and long before, social selection has spread royal and noble DNA much wider than the officially recognised genealogies, the lists of the great and the landed. For millennia powerful men often had a great deal of sex with many different women and, as has been observed, this cultural habit had spectacular results across central Asia through the efforts of Genghis Khan and his energetic brothers. As the population of Britain steadily increased after 1066 (despite catastrophes such as the Black Death), royal and noble DNA became significant. Scottish research supplies examples. More than 15% of all men who carry the surname of Stewart have S463, royal Stewart DNA, a sub-type of R1b-S145. The lineage of Robert the Bruce, the tough and dogged soldier who seized the throne of Scotland in the early fourteenth century, has also been found to be much more than an elaborate family tree. Around 10% of all men who carry the surname also match the DNA of two descendants of the Bruce with a sub-type of R1b-S116*, the king's marker. And Somerled, the first Lord of the Isles, who ruled in the twelfth century has more than 20,000 direct descendants living in Scotland now.

Royal and noble links reach across geography as well as history. Some are unexpected. A few decades before Gongu Hrolf met Charles the Simple at the monastery of Saint-Clair-sur-Epte, another band of Scandinavian warriors claimed extensive lands far to the east. Instead of sailing westoversea or down the North Sea coasts, Swedish Vikings penetrated the rivers of the Bay of Riga and the Gulf of Finland. By the end of the ninth century they controlled the river-borne overland route between the Baltic and the Black Sea. Known as Varangians, they established themselves on the Dnieper and even found employment further south, as imperial guards at Constantinople.

After 862, Rurik the Varangian founded the principality of Kievan Rus, the forerunner of Russia. It was based on the towns of Novgorod and Kiev, and this remarkable dynasty of Swedish Vikings succeeded in subjugating the Slavic tribes of the interior. Although Kievan Rus disintegrated and reformed, the lineage of the Varangians endured, and their DNA has been preserved. It is a sub-type of N-M46, labelled S431, and it was passed down to the early Tsars of Russia, including Ivan the Terrible.

In July 1991 a shallow grave was discovered at the site of a large house in the outskirts of Ekaterinburg, a busy industrial city in the foothills of the Ural mountains. Nine bodies were disinterred. There was strong evidence that they were the Romanovs, the skeletons of Tsar Nicholas II, the Tsarina Alexandra and their children. In the wake of the October Revolution, the royal family had been murdered on the orders of Lenin by Bolsheviks at Ekaterinburg – but were these bodies the Romanovs? DNA supplied answers.

The remains of Tsar Alexander's brother, the Grand Duke George, were exhumed and his DNA was found to be a perfect match for that of the only man buried at Ekaterinburg. Both men carried the mitochondrial marker T2a1a, and they had inherited it from their mother, the Tsarina Maria Feodorovna. Her name disguises her Danish origins and she was originally known as Princess Dagmar of Denmark of the House of Schleswig-Holstein-Sonderburg-Glucksburg, also known as the House of Oldenburg. Junior lineages stemming from the Oldenburgs include the royal families of Norway, Denmark, Greece and will also bring in that of Great Britain when Prince Charles, or his son William, becomes king.

The House of Schleswig-Holstein-Sonderburg-Glücksburg

More manageably known as the House of Oldenburg, this northern German ducal family stands at the centre of European royal genealogy. It is the senior branch of the extended family that includes the Danish, Greek, British (on the ascension of Prince Charles or Prince William to the throne – because Prince Philip is descended from the Greek royal house) and Norwegian dynasties. The title of His Highness the Prince of Schleswig-Holstein can only ever be His since it is passed on by agnatic primogeniture, allowing only males to succeed to the throne. Christoph, the current prince, has ensured that his line will continue since he has no fewer than five sons. This in turn means that Schleswig-Holstein Y-chromosome DNA is widely spread amongst European royal families.

This T2a1a sub-lineage reaches back many centuries to Elizabeth of Bohemia, who lived between 1409 and 1442. She is the ancestress in the female line of many European monarchs, 16 in all. Counted amongst them are four British kings; Charles I, George I, George III and George V. All of them carried the same mtDNA as Tsar Nicholas II and their most distant motherline ancestress was Adelheid von Alpeck, who died in Bavaria in 1280.

But these remarkably close connections did not end there. The mitochondrial DNA of the Tsarina Alexandra was also recovered and found to be a perfect match for that of Prince Philip, the Duke of Edinburgh. Both had sub-types of the haplogroup H, and both could trace their DNA back to Queen Victoria, who carried the same marker. It is H1af2, and she inherited it from her motherline which reached back to the lineages of German noblewomen. Anne of Bohemia and Hungary carried it. She lived between 1503 and 1546 and is the ancestress of no fewer than 36 European monarchs, all of whom shared her subtype. H1af2 was carried by four kings of Great Britain whose reigns spanned four centuries; Charles II, James II, William III and Edward VII. They can all trace their mtDNA back to a very distant female line ancestor: Mathilde de Provence, who was born in 1034.

A small number of DNA markers circulated amongst European royalty, which is deeply interlinked, with the same markers appearing time and again. But recent research has produced surprising connections. In the case of Prince Philip's grandson, Prince William, DNA has shown a startling new element in the royal lineages of Britain.

The story starts back when the East India Company was flourishing, and British men were making their names and their fortunes in India – and forming relationships with the women they met there. The Company was a phenomenon, a huge business enterprise that in effect ruled a subcontinent. It raised private armies, employed generals, fought pitched battles against other European colonists, founded towns and made vast fortunes for those involved. For young men hoping to make their way in the world of the burgeoning British Empire, it was one of the most exciting, dynamic and promising avenues to success. The Company traded in silk, cotton, opium, indigo dye, saltpetre and tea, and after its famous general, Robert Clive, defeated the French at Plassey in 1757, it ran India as a virtual monopoly. When Henry Dundas became president of the Board of Control in 1784, he began to appoint Scotsmen to key positions, so much so that by the end of the eighteenth century they dominated the activities of the East India Company, their connections reaching back to Scotland.

The estate of Boyndlie lies about five miles south-west of Fraserburgh in the north-east corner of Aberdeenshire. As the third son of John Forbes, Theodore will have known from boyhood that his future probably did not lie in farming. Some time in the early nineteenth century, he found himself in the Port of Leith working in a merchant company. Trade with India was brisk and Scottish entrepreneurs had invested so heavily in the tea industry that production there outstripped that of China. No doubt through contacts made in Leith, Theodore was promised a position with the

East India Company and he boarded a ship bound for the Bombay Presidency. India had been divided into three presidencies, or provinces, and the others were centred on Calcutta and Madras.

When Theodore arrived in India at the age of 21, he was posted to the port of Surat, about 200 kilometres north of Bombay. His agent was a man known as Aratoon Baldassier and he suggested that Forbes employ a housekeeper. She was Eliza Kewark, an Indian woman probably only two years younger, and Aratoon's sister-in-law. Her Christian name was almost certainly an anglicised version of Aleeza or Aliza. Not uncommon in the north-west of the subcontinent, it can mean 'Precious', or more prosaically, 'the daughter of Ali'. It may also have been a name popular with the significant Armenian community in India since her father appears in the historical record as Jacob Kevork or Hakob Kevorkian. Both versions sound very like typical Armenian names. No record survives of her mother, but Eliza's sister, the wife of Aratoon Baldassier, was called Khanay, a name with a definite Indian ring to it.

The thriving Armenian enclave in Surat had built a church where they could worship in their version of the Orthodox Christian faith. Researchers believe that Theodore Forbes and Eliza Kewark were married there, but that their union may not have been legally recognised by the British authorities. In any event, Theodore wrote in his notebook that his partner was 'the very pattern of what a wife ought to be' and in his letters he addressed her affectionately as 'My Dear Betsy'.

The Yemeni port of Mocha gave its name to a distinctive flavour of chocolate in Europe, perhaps because of its association with the coffee trade. Between the fifteenth and seventeenth centuries, there was a busy market in coffee beans at the old port. In the late eighteenth and early nineteenth centuries, the ships of the East India Company anchored in Mocha's harbour to take on fresh water and supplies and also engage in trade. It was an important staging post and some time after his marriage, Theodore was posted there. He and Eliza found themselves setting up home in Yemen, and in December 1812 they had a daughter, Katharine Scott Forbes. Known as Kitty, she was soon joined by a brother, Alexander, who was born two years later, in 1814. It seems that the couple's relationship was stable and settled, a genuine marriage, for they went on to have another child. Within a year, Eliza had returned to Surat where she gave birth to a second son, Fraser. But, very sadly, the little boy died aged only six months. At the same time it appears that Theodore had been offered a partnership in Forbes & Co, a busy trading company based in Bombay. The senior partner was a distant relative, Sir Charles Forbes. But attitudes were changing, and after more and more British women and wives came out to settle in India, relationships with native women began to be frowned upon. This shift in social mores persuaded Theodore, or his partners persuaded him, to leave his wife and two children in Surat.

A series of increasingly desperate letters survive. Probably dictated to a scribe (whose English was less than complete) by Eliza, they tell of the agonies of parting as she begged the father of their children to bring his family to Bombay, 'I entreat you my dear sir that you may call [us] from hence as soon as possible. Then [I] will be happy and [you will] save my life.' The letters were signed by Eliza in Armenian, and in earlier correspondence, she and Kitty wrote in Hindi. This suggests a mixed Armenian/Indian cultural background. In February 1818, Eliza wrote again asking for money and added 'Kitty and Alexander often ask after their beloved Papa and I let you know they are in good health'. But Theodore was about to let Eliza know that she would never see her children again.

In June 1818, Thomas Fraser, a friend of Theodore, wrote a letter that sounds like part of a longer exchange. He had visited Eliza and the children in Surat and commented, 'Kitty retains her good looks but the sooner you give the order about her departure to England the better, as her complexion will soil in this detestable climate'. Perhaps it was darkening too much under the Indian sun, too much for Kitty to be accepted back in Britain without the risk of being tainted as mixed race or 'of coloured blood'. Theodore decided that his six-year-old daughter should be sent back to Boyndlie in Aberdeenshire. Eliza must have been distraught but she insisted that the little girl be accompanied on the long voyage by Fazagool, her faithful servant. She wrote to Theodore, 'My good sir, I pray you let me know by your leave, I will bring my child to give in your hand by myself, and after Kitty is despatched to Europe then stay in Bombay or Surat.' This passage is from Eliza's last surviving letter to Theodore – perhaps they were reunited in Bombay. But it would not have been for long.

In 1820, Theodore Forbes decided to return to Britain. Having decided to leave Eliza behind, he boarded the SS *Blenden Hall* but died soon afterwards and was buried at sea. Perhaps he knew he was mortally ill, for he composed a will on board. In it he referred to Eliza as his housekeeper and left her a monthly allowance of only 100 Bombay rupees a month, less than half the amount she had been given before Kitty left for Scotland. To Kitty, described as his 'reputed natural daughter by Eliza Kewark', Forbes left 50,000 Bombay rupees. His 'reputed son', Alexander, was to have 20,000 rupees, but was commanded to stay in India. But it seems that Alexander disobeyed. There exists a record of a marriage of 'Alexander Forbes, bachelor, son of Theodore Forbes, merchant in Bombay (deceased) and Eliza Forbes MS Quark [*sic*] and Elizabeth Cobb in Dundee on 29th June, 1865'. At 51, Alexander was marrying late. Perhaps he had only recently returned from India. He and Elizabeth had two children and probably went to live in Arbroath. If Alexander had spent some time in India before returning to Scotland to be married and begin a family, he may have had knowledge of the family of his aunt and uncle, Aratoon Baldassier and his wife, Khanay. If they had had female children, the mtDNA carried by Eliza Kewark may well have lived on in India.

Alexander and Kitty certainly carried the mtDNA of their mother, and while Eliza's son could not pass it on, there is no doubt that that shared mtDNA lived on in Katharine Forbes and her descendants.

In the early nineteenth century and on into the Victorian age, illegitimacy was perhaps less of a stigma in the fermtouns of Scotland than it might have been in the genteel drawing rooms of the cities. Much more of a problem would have been the taint of 'coloured blood'. But since Katharine's father had died and her mother remained thousands of miles away in India, it may be that Eliza Kewark's ethnicity was not an immediate difficulty. Later, she was said to have been an Armenian, perhaps because Kewark or Kevork was recognised as an Armenian surname. Nevertheless, Eliza's existence was not forgotten or expunged from the family tree. Perhaps that was Katharine's doing, a stubborn unwillingness to deny her mother, the woman who had borne and raised her for at least six years in Surat. It is impossible to do more than guess at what was said and what was not.

In any event, Theodore and Eliza's daughter, also known as Kitty in Scotland, married James Crombie in Aberdeen. His family became well known for making excellent overcoats. Kitty was 25 years old and very striking looking – looks that were passed down the generations. Her family may have remained pillars of the Scottish middle class had Katharine's great-granddaughter, Ruth, not married into the aristocracy. Her husband was Maurice Burke Roche, 4th Baron Fermoy, an Irish peer. Ruth became a longstanding member of the household of Queen Elizabeth, the Queen Mother. In 1954 her daughter, Frances, married Edward, Viscount Althorp (later Earl Spencer) and in 1961 gave birth to a daughter, Diana Spencer. A year after her marriage to Prince Charles in 1981, she in turn gave birth to a son, Prince William. In the direct female line, Eliza Kewark's mitochondrial DNA had been passed down to the heir second in line to the throne of Great Britain and Northern Ireland.

How is it possible to be certain of this? Mitochondrial DNA is passed down the motherline to all children. Two living direct descendants of Eliza Kewark have been found and by reading the sequence of their mtDNA, geneticists discovered not only that it matched but that it also belonged to a haplogroup called R30b. Further research confirmed unequivocally that this was Eliza Kewark's haplogroup. A comparison run through databases of the DNA of more than 65,000 individuals from around the world showed that R30b is very rare and very Indian. Only 14 examples have been reported and 13 of these were Indian, with one in Nepal. To add to this research, it is important to note that the other related branches of R30b, that is R30a and R30, are also entirely south Asian in origin. This confirms beyond doubt that the mtDNA of Eliza Kewark was of Indian heritage.

R30b is rare even in India, where only approximately 0.3% of people carry the lineage. And what Eliza passed down to Princess Diana and Prince William, and her other living descendants, is even rarer. Within the haplogroup of R30b, an exact match to her sequence has yet to be found outside of her descendants. But Prince William, and Prince Harry, who also carries it, will not be able to pass on their extremely rare Indian mtDNA to their children, who will inherit whatever their mothers' mtDNA happens to be.

For yet more corroboration, scientists used an independent type of genetic evidence. By reading over 700,000 markers scattered across the genome of Princess Diana's matrilineal cousins, and comparing findings to a global database of samples, it is possible to estimate the proportions of continental-level ancestry for an individual. For example, someone with a father from Ireland and a mother from Nigeria would be 50% sub-Saharan African and 50% European, or someone with three English grandparents and one from China would be approximately 20% to 30% east Asian. The proportions inherited from ancestors who lived longer ago are lower and also variable. Eliza Kewark's two royal descendants are estimated to be about 0.3% and 0.8% south Asian, with three blocks of south Asian DNA in each of their genomes. All of the rest is of European origin.

It is therefore very likely that in addition to his mtDNA, Prince William has not only inherited a small proportion of Indian DNA from Eliza Kewark but that his heirs will also carry it. In a moving footnote, Princess Diana's brother, Earl Spencer, spoke with great feeling at his sister's funeral of his and his nephew's blood family. His own daughters are called Eliza and Kitty.

Scipio Kennedy

On 6 February 1725 Scipio Kennedy, a black slave from Guinea on the west coast of Africa, was given his freedom at Culzean Castle by Sir John Kennedy. It was fashionable in Scotland and elsewhere to own a black slave or page boy, something of a status symbol. Scipio stayed in service to the Kennedys, taking their name but also being paid. Two years after being granted his freedom, the young man was accused of fornication with a local woman, Margaret Gray. He later married her and they had a daughter, Sarah. She and their other children all took the name of Kennedy. The genetic consequence of Sir John's humanitarian gesture in 1725 is that there is now west African DNA in the heart of Robert Burns' homeland, and because of the Scottish involvement in the slave trade, there is likely to be more.

At least one observation emerges from even a brief overview of royal and noble DNA throughout history and this particular, more modern example. Despite the limited number of markers shared by modern European kings and queens, their DNA is not exclusive. Over the span of the ten centuries since the Norman Conquest, dynasties have absorbed new and sometimes unexpected lineages. Far from being a closed elite insulated from the rest of society, royal and noble families have spun a web of links across social and economic barriers and across millennia. The succession of kings and queens may appear to be a very traditional, even limited framework for a history of Britain, but many more people can identify with the fortunes of dynasties, even if they were unaware until recently of the connections made by their DNA.

The links forged by William the Conqueror were with France, but his ambitions in the mainland of Britain did not diminish. Once he had crushed native resistance in the ruthless campaign known as the Harrying of the North, the new king of England turned his attention to the west. In 1081 he made what was either a pilgrimage or a reconnaissance, or both, to the shrine of St David in Pembrokeshire. There he met Rhys apTewdwr, Prince of Deheubarth. His Tudor patronymic is not ancestral to William I's distant successors but it is interesting to see the Welsh version of Theodoric survive so vigorously. Relations soured a decade later as the Conqueror's son, William Rufus, and his powerful barons attacked Wales and Rhys apTewdwr was killed in 1093. His death laid Deheubarth open to Norman colonisation and a clear distinction quickly grew between Wallia Pura, or Welsh Wales and Marchia Walliae, the borderlands of Wales, the territory occupied by Norman lords. And with the arrival of Flemish and French colonists, Pembrokeshire changed character to become 'Little England beyond Wales'.

The Norman kings were a vicious dynasty, unhesitating in pursuit of personal ambition, particularly when it came to siblings and other members of their own family. William the Conqueror died in 1087 from battle wounds. His son and impatient heir, Robert Curthose, had made an alliance with Philip I, King of France, to eject his father from the duchy of Normandy. It was seen as the more important possession and as eldest son, Robert became duke while his younger brother, William Rufus, succeeded to the throne of England. When in 1100 William was killed while out hunting, his younger brother Henry seized both the English crown and the Norman duchy. Robert Curthose was on crusade but when he returned he not only recovered Normandy but invaded England. Family business was finally resolved in 1106 at the Battle of Tinchebrai with Robert's defeat. Henry I once again made himself duke of Normandy.

In England the Norman elite appears to have had little interaction even with the upper echelons of native English society. They may not have understood each other. The new masters spoke French and wrote in Latin. Normans seldom married native women. But despite this clear and continuing cultural separation, the tiny ruling minority began to be influential in surprising ways.

Not only did surnames become more common after the later eleventh century, the fashion for Christian names changed. Up until the Conquest, children had been given Anglo-Saxon and Danish names and these could be tremendously varied. In Devon alone 562 different Christian names were recorded before 1066. But under the Normans biblical and saints' names grew tremendously in popularity, as did those of the conquerors themselves. William and Henry were especially favoured and the only significant survivor from the past was the name of Edward. Choice appeared to shrink markedly, and that underlined the need for surnames so that identity could be made clear.

As immigrants continued to cross the Channel, new names indicated their origins: Flemings came from Flanders, Burgoynes from Burgundy and Bremners and Brabhams from Brabant in Belgium. In Scotland, Bremners penetrated as far north as Wick and the 1881 census shows 642 people of that name living in the county of Caithness. This concentration may have much later links with the herring fishing and the trade with the Low Countries.

Surnames such as these are probably as reliable an indicator of origin as DNA in these particular cases, since the frequency and distribution of markers in northern France, Belgium, Holland, northern Germany and Denmark is similar to that of southern and eastern England. Clear patterns are difficult to discern.

William the Conqueror did bring a group of people to England whose origins were much more distant and many of whom will have carried distinctive markers. William of Malmesbury reported that Jews came from Rouen in the late eleventh century, probably to settle initially in London. What motivated this migration was almost certainly money. William I wanted his tenants not to pay him in the traditional way with foodrents, goods and services but to settle what they owed in cash. And there was insufficient access to coin in England after 1066. In essence the Jewish communities were bankers, and in order to supply cash in return for pledges or indeed valuables, they were allowed to travel without hindrance. No tolls would be levied on Jews and they were granted legal privileges. Great weight was given to the oath of a Jew. It was equivalent to an oath sworn by 12 Christians.

The anti-semitism stirred up throughout Europe as a consequence of the crusades spilled into England. What was known as the blood libel began to circulate, the notion that Jews used the blood of Christian children in rituals, and King Stephen, the successor of Henry I, had to act to prevent attacks on them. Jewish communities grew up across the south of England, from Canterbury to Gloucester, and it was said that in the twelfth century the richest man in England, richer even than the king, was Aaron of Lincoln. By the end of the thirteenth century resentment had boiled over into persecution and massacre. When Edward I enacted a statute that forbade usury, as the lending of money with interest was known, he effectively put the Jewish community out of business. Those who remained were finally expelled in 1290, and until the middle of the seventeenth century, very few Jews lived in England and there appears to have been no community.

The common broom with its brilliant yellow flowers seems to have been the badge of Geoffrey, Count of Anjou. Worn before the age of uniforms to identify his supporters and soldiers, it had the Latin term of *planta genista*, a name that became famous as Plantagenet. Geoffrey's son Henry succeeded Stephen in 1154 as king of England and the ruler of a vast Anglo-French empire that stretched from the Pyrenees to the Cheviot Hills. Henry II was the most powerful man in Europe. Constantly travelling, ferociously defending his interests and attempting to expand them, he consolidated the border with Scotland, taking back Cumbria from the macMalcolm kings. But this restless king spent much more of his time in France. In a long reign of 34 years, he spent only 13 in England. After Henry's death in 1189 his son Richard ('the Lionheart') went on crusade and under him and his brother John, the Plantagenet empire began to disintegrate.

Nevertheless, English kings remained determined to retain their French possessions. For the best part of three centuries their armies fought in France, and the last shred of the vast territory held by Henry II was lost in 1558 with the fall of Calais during the reign of Mary Tudor. The Plantagenets were more successful in Britain. Edward I colonised Wales, and Ireland fell under English control. Only Scotland retained her independence, and then only with difficulty.

In the middle of the fourteenth century, England's French connection proved disastrously fatal. In June 1348, a ship set sail from Gascony. By the time it weighed anchor at Weymouth in Dorset, one of its crew had developed tell-tale and terrifying symptoms. The Black Death had made landfall. Carried by the fleas on black rats, its impact was devastating, killing between 75 and 100 million people in Europe at its height. Estimates vary, but the consensus is that the population of medieval England was about 6 million before the arrival of the plague, and it may be that between 30% and 50% died as it raged through communities. Its impact on Britain's ancestral DNA is very difficult to measure. If entire lineages were wiped out by the Black Death and consequently can no longer be found in the modern population, then part of DNA history is truly lost.

What can still be found is ancient DNA, and it can make for sensational, fascinating headlines. In February 2013, scientists at Leicester University confirmed that they had retrieved the mitochondrial DNA of the last Plantagenet king, Richard III. He was famously killed in 1485, in battle with Henry Tudor, the man who would become Henry VII.

Richard III was not only the last Plantagenet but also the last English king to be killed in battle, and his body had never been found. The phrase 'the king is dead, long live the king' was sometimes more than a formula and a matter of important and emphatic show. It was very unusual for a royal corpse simply to be disposed of, with no record or marking of its burial. Bosworth Field, the place where

Shakespeare had Richard roar 'My kingdom for a horse!' is only a short distance from Leicester, and near-contemporary sources did record that the king's body had been brought there, to the Church of the Grey Friars, for burial. As early as 1986 historians had been suggesting that the king's body had been buried in the choir of the church. After the dissolution of the monasteries in the reign of Henry VIII, it had been demolished and built over. Part of the site is now a car park for Leicester University's department of social sciences. When Philippa Langley of the Richard III Society became involved, momentum in the search for the king's grave began to build. In 2009 she approached a group of archaeologists who had been associated with Leicester University, and in 2012 a dig began. After piercing the tarmac and locating the foundations of the old church, a body was exhumed. Activity intensified and expectation escalated.

An osteologist from Leicester University meticulously examined the skeleton and it showed all the brutal marks of having been cut down and killed with bladed weapons. There was clear evidence of ten serious wounds, eight to the head and two to the body. It seemed that this soldier had indeed fought on foot, supporting the contention that this was Richard III, who had fought hand to hand in the thick of battle. Even the most damning Tudor propaganda did not deny his physical courage. A large part of the base of his skull had been sheared off by a blow from a sharp blade, probably a sword, and on another part of the head there was the mark of a severe puncture wound, perhaps made by the spike of a halberd. Either would have been fatal. The remaining wounds to the head were in the face, what are known as humiliation wounds, and inflicted after death or while he lay dying. And when his armour was stripped off, the body was once again stabbed. A blade was thrust into the buttock, possibly when the corpse had been slung over the saddle of a pack-horse to be led from Bosworth to Leicester. This interpretation of events on the battlefield is made more likely by the fact that the wrists of the skeleton were found to be crossed when it was exhumed. The hands had probably been tied and the body secured to the pack-saddle.

Radiocarbon dating added more support to the growing sense that the body in the car park was indeed Richard Plantagenet. When the bones were tested they indicated a date range between 1450 and 1540 and an age between 30 and 32. All of which fitted. The Shakespearean stereotype of Richard Crouchback also began to emerge from the shadows of myth-history. Rather than a fiction, or indeed a monster, created by later writers, the skeleton turned out to be that of a man who suffered from scoliosis, a condition that meant one shoulder was higher than the other. Near-contemporary sources also talked of Richard's slender and almost feminine physique, and the skeleton in the car park confirmed that description.

The evidence mounted, but it was to be DNA that would finally prove the case beyond all reasonable doubt. Mitochondrial DNA was carefully extracted from a tooth and then analysed. Richard III had no surviving legitimate children, or at least none that were recorded. That meant genealogy had to be brought into play. Richard shared his mother Cecily Neville's mtDNA with his sister, Anne of York. Crucially, she could pass it on to her children. If an unbroken and direct line of matrilineal descent could be securely identified, then mtDNA from a living descendant could be extracted and analysed. And more, if it matched what had been recovered from the skeleton, then it really was Richard III.

In 2003, the historian John Ashdown-Hill had traced what he believed to be an unbroken line from Anne of York. Her mtDNA was carried by a retired journalist living in Canada, Mrs Joy Ibsen. She died in 2008 but her son, Michael, had inherited her DNA and he supplied a sample. It was compared to that of the skeleton buried at the Grey Friars – and it matched. But for further corroboration, Leicester University traced a second living descendant of Cecily Neville, someone who wished to remain anonymous, and this person's mtDNA provided a clinching second match. To make all even more secure, it turned out that Richard III's marker was a rare sub-type of the mtDNA haplogroup of J.

This remarkable project did more than solve the mystery of what happened to the last Plantagenet, it also saw DNA used as a central element of evidence not so much in rewriting history as writing better history.

If Richard III roared for a horse on Bosworth Field, he almost certainly roared in English. Richard II was the last king of England whose first language was the Norman French of his ancestors. By the fifteenth century English had replaced it in the law courts and other areas of royal and local administration. While Latin was reserved for the Church and written record, Norman French had marked out a clearly defined social elite for 350 years. It left a substantial inheritance with everyday words such as cabbage, candle, fork, beef, pork and many others. While Parisian French replaced the dialect of William the Conqueror and the Plantagenets across mainland France, Norman French was almost completely effaced, except in one place. It is still spoken in the sole remnant of the once vast French domains of English kings, the Channel Islands. And at Westminster, echoes of the medieval past are heard when bills pass through Parliament. Formal phrases in Norman French are still used for stages such as royal assent.

When Henry Tudor defeated Richard III, he will have issued orders in English but may have spoken to the captains of his Welsh companies in their native tongue. His grandfather, Owen Tudor, certainly spoke Welsh and since both of his sons held lands in Wales, it is likely that Henry's father, Edmund, had the language despite his English name (Owen took his grandfather's surname and not his father's. If he had, the dynasty would have been called the Merediths). In 1485 eight languages were spoken in Britain. In addition to English, Norman French and Welsh, Cornish, Manx, Irish, Scots Gaelic and Norn all had significant speech communities. Across about half of the landmass of the British Isles and Ireland, Celtic languages were heard and about a third of the entire population spoke one of them. North of the Trent and in Lowland Scotland, an essentially Northumbrian dialect of English was understood and in the south of England, Midland dialects were spoken. The south-east was yet to become influential. This patchwork survived surprisingly late and was kept in place largely because of lack of mobility. Communities were born, lived and died inside a restricted radius, and this relatively unchanging scene must also have had the effect of pinning particular DNA markers to the map.

French came back to Britain in the later sixteenth century, not in the mouths of conquerors but rather those who were fleeing from appalling persecution. On 24 August 1572, Paris should have been celebrating a royal wedding. Henry of Navarre had married Margaret de Valois, sister of the French king Charles IX, but instead of bunting and pageantry, the streets of the city were bathed in blood. Henry was a Protestant who had declined to have his marriage consecrated by a mass at Notre Dame, and the Paris mob was fervently, violently Catholic. Probably at the instigation of Catherine de Medici, the king's (and Margaret's) mother , what developed into the St Bartholomew's Day Massacre began. Many of those Protestants who had come to the city to see the royal marriage were slaughtered in the streets, perhaps 3,000 in Paris alone. The killing spread to other towns and cities and it is estimated that a further 10,000 were murdered.

In France Protestants were known as Huguenots, and after the shocking events of August 1572 many of them fled abroad. Once again persecution by a Catholic government drove many across the Channel. It is believed that by 1573 there were as many as 10,000 refugees in England. And they kept coming. Oppressed by Cardinal Richelieu and later by Louis XIV, there were more than 23,000 Huguenots in London alone by 1700.

These immigrants were often welcomed for they were skilled, middle-class workers. France had a population of about 20 million in the 1690s compared with only 6 million in Britain. Considered wealthier and more cosmopolitan, French society had provided a market for a range of skills and trades practised by Huguenots; the likes of clock-making, gold- and silversmithing, paper-making and above all creating fine textiles. When they arrived in the south of England, the religious refugees set up as weavers and retailers in London and several other English towns.

The ancestral heritage of this immigration is better observed in surnames than in DNA markers. Not only did they add to the already mixed genetic picture of the south-east, the markers the Huguenots brought were often those of north-west Europe, many of which were already well established in England. But names, sometimes famous, like Olivier, Courtauld, Batchelor, de la Mare, Garrick and Bosanquet, do recall the coming of what was a significantly large group of new people to Britain, Beyond the south of England, there were very few Huguenot communities: just one in Chester and another in Edinburgh. The French surname of Dumma is still occasionally found in Scotland.

Lots in a Name

A recently compiled surname map of Britain used a combination of data from the electoral roll of 2007 and Twitter use. Across the nation names such as Smith, Brown, Jones and Williams tended to dominate, with more unusual names appearing on the periphery: MacLeods in Lewis, Rowes in Cornwall and Prices in Wales. In London the most striking finding was that Patel rivalled Smith as the most common surname. Since men inherit surnames with their Y chromosomes and keep them, these patterns can be important for looking at how the distribution of our collective DNA is shifting.

In 1688, 100 years after the Huguenot French influx to England, the feckless Stewart dynasty fled Britain for France. The birth of a son to James VII and II in June of that year had proved a dramatic catalyst, because it converted the much-feared possibility of a Catholic succession to the throne into a likelihood. A group of Whig and Tory politicians invited William of Orange to intervene and take the throne. He was married to Mary, the Protestant daughter of James VII and II, and when he landed at Brixham in Devon in November 1688, the Stewart king sought sanctuary in France some weeks later.

James's dynasty retained support in Scotland and the succession of William and Mary set in train a decisive period in the history of the Highlands, where James' supporters became known as Jacobites. Until 1746, Jacobitism was backed by armies of clansmen and it represented a powerful political alternative. But after their calamitous defeat at Culloden, a policy of genocidal government repression began to empty the glens and straths of their people. Accelerated by the actions of a brutal aristocracy, migrants were increasingly forced to move south to the cities of the central belt of Scotland or they took ship for the colonies in North America and the southern hemisphere. These migrations took place over the following 200 years. Most did better than survive and many prospered. The characteristic DNA markers of the clans are consequently far more numerous overseas than they are in Scotland.

When Queen Anne died in 1714, there were no more Protestant Stewarts left. But succession planning had begun after the death of her predecessor, William III (aka William of Orange), in 1702. Georg Ludwig von Braunschweig-Lüneburg, the Duke and Elector of Hanover in Germany, was a distant relative, the son of Queen Anne's cousin by several removes, Sophia. Famously ill-tempered, he had divorced his wife and imprisoned her in a castle for life, and he was not on speaking terms with his son and heir, Georg Augustus. But he was a Protestant. On 1 August 1714, George I was proclaimed King of Great Britain and Ireland, and for the sake of completeness, he also awarded himself the non-existent title of King of Hanover.

George was descended from an ancient line, from Guelf IV, the Duke of Bavaria in the eleventh century,which appears to have come down unbroken until the death of William IV in 1837. The throne then passed to Victoria, the daughter of Edward, Duke of Kent, the late king's younger brother. Both the Tsarina Alexandra and Prince Philip, the Duke of Edinburgh, are directly descended from her in the female line. They carry the H1af2 mtDNA marker, which they must have inherited from Victoria and which she passed on in other directions. Because she and Prince Albert had nine children and 42 grandchildren (not all of whom survived into adulthood), that means that the H1af2 marker passed down to Victoria's five daughters is now very widespread amongst the remaining European royal families and nobility. H1 is in any case one of the most common European mtDNA haplogroups.

By contrast, Prince Albert's Y-chromosome DNA marker is rare. He carried R1b-S376 and passed it to Edward VII, George V, Edward VIII and George VI. Since the last of the Hanover Georges (by this time diplomatically renamed Windsor) did not have sons, Queen Elizabeth II ascended the throne on his death in 1952. She carries the mtDNA of her mother, Elizabeth Bowes-Lyon and has passed it to Princess Anne and Anne's daughter, Zara Phillips. But the Y-DNA marker of Albert, R1b-S376, has not died out. Directly descended from George V in the male line, Prince Michael of Kent, the Duke of Kent and the Duke of Gloucester all carry it and have passed on the royal marker to their sons and grandsons.

Prince Philip is not only a direct descendant of Queen Victoria and a relative of Tsarina Alexandra through his motherline, he may also carry the Y-chromosome marker of Tsar Nicholas. When bones were exhumed from the hurriedly dug graves at Ekaterinburg in 1991, Prince Philip gave both blood and hair samples to see if his mtDNA was that of the Tsarina. It was a perfect match. Partly because techniques were less advanced, the Y-chromosome DNA of Tsar Nicholas was not investigated until later. In 2009 it was reported that Y-DNA had been extracted from his skeleton, but a confirming match was needed. In 1891 Nicholas II was attacked in Japan in what appeared to be an assassination attempt. An escorting policeman suddenly swung his sabre at the tsar's head and cut him badly before being fought off by Prince Georg of Greece. The bloodstained shirt of the injured Nicholas II had been preserved as a relic and Y-DNA was successfully extracted and analysed. It matched the Y-DNA taken from the bones found at Ekaterinburg.

It turned out that the tsar and his defender, Prince Georg of Greece, were probably related, both directly descended from Frederick I of Denmark in the male line. Providing that there were no illegitimacies over 29 generations, that means they shared the same Y-chromosome marker. Before his marriage the Duke of Edinburgh was known as Prince Philip of Greece, and he is also a direct descendant in the male line from Frederick I, King of Denmark. All of these men probably carry or carried R1b, the most common Y-DNA haplotype in Europe.

Through Prince Charles, the R1b marker has likely been passed on to Prince William. That potentially makes his mix of mtDNA and Y-DNA very contrasting. On the one side he carries the extremely rare Indian marker he has inherited from Eliza Kewark while from his father, he has the same Y-DNA haplogroup carried by 60% of all European men.

10

Comings and Goings

✴

E WEN MULLINS IS AN Irish plant geneticist at war with the past. In his research facility outside Carlow, he battles against potato blight, *Phytophthora infestans*, by genetically modifying plants so that they become resistant to the disease. His work is the latest skirmish in a war that has been fought in the fields of Ireland for 128 years. Potato blight was first seen in Flanders as the crop began to rot instead of ripen. The disease spread rapidly to Britain and Ireland and in 1845 and 1846 the potato harvest failed. In the devastating famine known in Gaelic as An Gorta Mor, the Great Hunger, approximately 1 million Irish men, women and children died. It was a turning moment. Between 1841 and 1851 Ireland's population fell from about 8 million to 6.5 million and for the next 100 years it kept on falling, reaching a low of 4.3 million after the Second World War.

An Gorta Mor triggered waves of emigration, 'as though people were fleeing from a burning building' according to one historian, to Britain and further afield, principally to North America. The movement of Irish people, driven from their homes by hunger and poverty, was significant, the numbers large enough to represent the most dynamic demographic shift since the early Middle Ages. Relatively rapid and cheap transport effected dramatic change. Between January and April in 1848, 42,860 Irish people disembarked at the quays on Clydeside, and larger numbers docked at Liverpool in the same period. After Christmas 1848, it was reported by a local journalist that huge numbers came to Merseyside, most of them en route for America and other overseas destinations:

The arrivals have amounted to nearly 300,000 and of these I believe the number now located among us in addition to our ordinary population is very moderately estimated at from 60,000 to 80,000 . . . they have in many instances been found sleeping in privies and even in the open street.

These two British cities became and remain the most heavily influenced by Irish immigration. They came to Liverpool, Glasgow and elsewhere and they kept coming, principally from the south and west, from counties Cork, Kerry, Tipperary, Limerick, Mayo, Galway and Sligo. Many also made the short voyage across the North Channel from Ulster to Scotland, but not all flocked to the cities and their factories and forges. By 1851 rural Wigtownshire had a large Irish population, 16.5% of the county's total, and working mainly in farming.

Of course, the refugees brought their characteristic DNA markers with them, but because Irish immigration – admittedly in much smaller numbers – had been going on for many centuries, it is difficult to identify new arrivals. The Y-chromosome marker of Niall Noigiallach, R1b-M222, crossed the North Channel in the fifth and sixth centuries AD to found the Gaelic-speaking kingdom of Dalriada, but there can be little doubt that a second wave burst on Scotland's shores after the Great Hunger. Its frequency is 1.8% in England, and while some of that may be from those earlier immigrants or brought south from Scotland, many men carrying M222 came after 1845. This surge in immigration drove up the numbers of Irish-born immigrants to Scotland to 7.2% of the total population, while in England and Wales it doubled to 602,000 or 3% of the total by 1861.

The human disaster of the famine and those it displaced was part of a developing pattern of migration that began to gather momentum in the nineteenth century. Carried on mass sea and land transport, men, women and their families began to move, sometimes far from their origins. Between 1821 and 1911, the population of England and Wales increased very rapidly as cities grew and the industrial and agrarian revolutions took effect. It tripled from 12 million to 36 million. Over the same period the population of Scotland also grew, but more slowly, from 2.1 million to 4.8 million. And between

1841 and 1911, the number living in Ireland fell catastrophically by almost half, from 8.2 million to 4.4 million.

In the later nineteenth century migration and emigration had profound and paradoxical effects. Despite substantial immigration from Ireland, the population of Scotland grew less dramatically because many Scots migrated to England. And after the ruthlessness of the Highland Clearances, when landowners forced their long-standing tenant smallholders out, many also sailed for new lives in the Americas, Australia, New Zealand and elsewhere. Nevertheless, by far the largest percentage of native British to settle in Canada, South Africa, Australia and New Zealand were born in England and Wales. And only Irish emigration to the USA outstripped the numbers of English and Welsh people who made new homes there. In all, more than 17 million people from Britain and Ireland emigrated between 1821 and 1911. After a late wave of departures following the First World War, numbers declined sharply. Almost a century of emigration from Britain was over. It was once more time for new people to come.

The brute effects of politics as well as the hardships of famine caused surges of immigration from another direction towards the end of the nineteenth century. In 1791 Catherine the Great of Russia established what became known as the Jewish Pale, a large area that included parts of Lithuania, Poland, Belarus, Moldova, Ukraine and western Russia. It was intended as a clearly determined area where Jews could live and work unhindered – and be taxed by Catherine's administration. It was also a clear target and occasionally anti-semitism burst into flames as mobs attacked the towns or shtetls. From 1881 onwards the riots and violence known as pogroms became more sustained as Jewish houses and businesses were burned in their thousands. Many died, and immediate emigration began. Most sought sanctuary in what they called 'di goldene Medine', the Golden Land of the USA, but because they fled first to Baltic ports, families found themselves disembarking on the eastern coasts of Britain, a place of temporary safety. Many stayed and the Jewish community in England of 46,000 in 1880 rose spectacularly to more than 250,000 by 1919. In response to anti-semitic pressure, the government introduced legislation in 1905 to restrict immigration. However, it applied only in England and Wales and the Jewish community in Scotland continued to grow.

The Anglesey Bone Setters

In the middle of the 18th century a mystery washed ashore on Anglesey. Two boys were the sole survivors of a shipwreck but since neither spoke a word of English or Welsh, they could not explain what had happened or where they came from. One boy died soon after his rescue, but the other, named Evan Thomas, was taken in by the local doctor. Because of his dark skin, Evan was thought to be Spanish. Having taken an interest in his adoptive father's medical work, the boy began to exhibit a remarkable skill. Working at first with animals, he used touch alone to feel where the creature had broken a bone and he was able to re-set it. Human patients soon appreciated his ability to manipulate the edges of a fracture so that they knitted better. Evan's techniques formed the basis of modern orthopaedic surgery. Remarkably, he founded a dynasty of bone-setters, eight generations of men who could heal broken bones. In the late 19th century, Hugh Owen Thomas invented the Thomas splint, which greatly reduced deaths from femoral fractures, and his nephew, Sir Robert Jones, used it to bring down fracture deaths on the Western Front from 80% in 1916 to 8% by the end of the First World War. In the fascinating question of where Evan Thomas and his amazing skills came from, DNA took a hand. He came not from Spain but eastern Europe.

In the late 1930s large numbers of refugees from Nazi persecution arrived in Britain, 40,000 Jews from Germany and Austria and around 50,000 from Poland, Italy and elsewhere. This made for a substantial and clearly culturally defined group. Near-Eastern markers can be distinguished in the modern population of Britain but specifically Jewish DNA is hard to find. Nevertheless, there exists an interesting phenomenon known as the Cohen Modal Haplotype. This is a Y-chromosome marker in the P58 group which is highly enriched amongst the Cohanim, the Jewish High Priests, men who often have the surname of Cohen. It confirms that a very large proportion descend from one man who lived about 2,000 years ago. According to tradition, this may have been Aaron, the older brother of Moses.

By the outset of the twenty-first century, the Jewish community in Britain was much reduced. Only a remnant of 6,400 remain in Scotland. Emigration to Israel after 1945 was one factor and another the growing tendency of Jews to marry outside their faith and culture, and therefore re-define themselves as non-Jews. It is one facet of a relatively rapid process of integration, common enough amongst immigrant groups.

As cities grew and social life spilled out of overcrowded housing, street food began to be popular in late Victorian Britain. Cockles, winkles and other forms of cheap seafood that were sold from baskets, and treats like roast chestnuts, were popular and well-established British traditions but what helped revolutionise outdoor eating was the arrival of Italian immigrants. Poverty was what drove most of them from their native home. In the 1870s families came from the Liri Valley, north of Naples, and they brought with them the skills for making ice cream. Britain's markedly colder climate no doubt helped it keep. Italians invented the pokey hat or ice cream cornet and as they sold them from barrows or carts, they made the whole retail operation simple and unfussy in that everything could be eaten.

Fish and chips appears to have been pioneered by Jewish immigrants to East London but when Italians began to open cafés and ice cream parlours, they often had fryers installed. Little Italy in Clerkenwell was the first focus of settlement but small groups soon spread around Britain. Part of the reason for such a wide dispersal was the Italian talent for street food and cafés. Not wishing to compete with each other, they often learned skills in one town and then moved on to another to open a chip shop, ice cream parlour or café. Following the Mediterranean tradition (albeit without the accompanying climate) of the evening walk and outdoor socialising, Italians kept their premises open much longer than the native British competition. As a result, cafés became a popular alternative to pubs and somewhere women and younger people could feel more comfortable.

Numbers were small, with the largest Italian community amounting to no more than 50,000 in London and 35,000 in and around Glasgow by the late 1930s. The immigrants added colourful and affordable diversity to Britain's high streets and they were generally popular. But in 1940 Mussolini changed attitudes abruptly when he declared war on Britain as an ally of Nazi Germany. After Winston Churchill made a speech about enemy aliens living in Britain and urged the police 'to collar the lot', thousands of Italian men between the ages of 17 and 60 were imprisoned. The actor Tom Conti's father was amongst them. A hairdresser in Paisley, he was arrested and interned on the Isle of Man. It was a pointless exercise, but perhaps it protected some since anti-Italian rioting broke out in June 1940 when Mussolini joined Hitler and many shop and café windows were smashed.

Italians not only added to Britain's social and cultural life, they also brought DNA that was sufficiently different as to be noticeable. When Tom Conti had his analysed, the Y-chromosome marker he inherited from his father, Alfonso, turned out to be very different indeed. Labelled E-M34 and described as Saracen, it originated in the Near East. In the eighth and ninth centuries Saracen pirates raided the coasts of the Tyrrhenian Sea and established a series of enclaves. Tom's father's family lived for many generations in coastal Lazio and then much later brought the marker to Scotland. It is very rare, with only 0.4% of men carrying it in England and 0.1% in Scotland. E-M34 also travelled to a part of France and it turned out that Tom Conti shared his rare marker with the most famous Frenchman in history.

From the tiny scraps of skin attached to beard hairs that had been preserved in a reliquary, researchers managed to sequence the Y-chromosome DNA of Napoleon Bonaparte. To be certain, they then tested that of his direct male-line descendant, Prince Louis Napoleon Bonaparte, the French politician who is the great great grandnephew of the emperor. They matched. And so did Tom Conti's DNA marker.

Napoleon's family was originally Italian and could trace lineages back to the tenth century and the coasts of the northern Tyrrhenian Sea. Much later, Giovanni Buonaparte settled a branch of the family in Corsica, which in turn became part of pre-revolutionary France. What these fascinating and unlikely comparisons mean is something simple – that the Conti and Bonaparte fatherlines descend from a relatively recent common ancestor whose DNA probably derived from the Saracen pirates of the Near East. All of which adds to Britain's genetic diversity.

More Italians came in the 1950s and through a quirk of industrial recruitment, many settled in two English towns. The London Brick Company (based in Peterborough and Bedford) preferred to advertise for workers in the southern Italian provinces of Calabria and Puglia and by 1960 more than 7,500 men were living in Bedford and 3,500 in Peterborough. An Italian vice-consulate was set up in Bedford. Now one in five of its inhabitants are of recent Italian extraction, an immigrant population of 14,000 living in a relatively small town. While this made for entertaining differences of allegiances every four years as the football World Cup came around, these local spats were a forgotten sideshow as Britain became a destination for much larger groups of immigrants from around the world.

On the evening of 20 April 1968, Enoch Powell MP made an infamous speech to a Conservative Association in Birmingham. He held the nearby seat of Wolverhampton South West and began his remarks by recounting the details of a recent conversation with a constituent. Described by Powell as a decent, ordinary, fellow Englishman, he wanted to see his children settled overseas because 'in this country in 15 or 20 years time the black man will have the whip hand over the white man'. As he warmed to what he knew full well would be a widely reported speech (his office had circulated advance copies to the press), Powell went on to predict that in 15 or 20 years there would be 3.5 million Commonwealth immigrants and their descendants living permanently in Britain. And by the year 2000, there would be more, between 5 million and 7 million immigrants, a tenth of the entire population. And then his instincts for the headline and those of a classical scholar combined to prompt this quote from Virgil's *Aeneid*: 'As I look ahead, I am filled with foreboding; like the Roman I seem to see "the River Tiber foaming with much blood".'

As he knew it would, Powell's speech and his 'rivers of blood' quote electrified opinion and made him into a major public figure. But it almost certainly had the opposite effect to what he intended, even though his predictions of the numbers of immigrants turned out not to be wildly wrong.

257

After the Second World War, the British Empire began quickly to disintegrate. India and Pakistan gained independence in 1947, and beginning with the election of Kwame Nkrumah as Prime Minister of Ghana a decade later, the African colonies followed suit. By the 1970s what the Labour Prime Minister, Harold Wilson, called East of Suez had gone and in the 1980s the smaller Caribbean islands and Zimbabwe were independent. Only a scatter of tiny dependencies and the disputed territories of the Falkland Islands and Gibraltar were left. It was a remarkably rapid and radical period of transition.

As a consolation prize and a means of maintaining British influence, the Commonwealth became more important and the queen remained a titular head of state in several member countries. Legislation had been enacted to maintain the sense of a community of nations and in 1948 the British Nationality Act allowed around 800 million subjects in the disintegrating British Empire the right to live and work in Britain without the need for a visa. Immigration began to rise from around 3000 a year in 1953 to 136,400 in 1961. Many were actively encouraged. Jamaicans and Trinidadians in particular easily found jobs in transport and in the National Health Service. But as the Empire changed into the Commonwealth, politicians began to become anxious. In 1961, the Home Secretary, Rab Butler MP, introduced new qualifications into the Commonwealth Immigrants Act by making a simple point, 'a sizeable part of the entire population of the Earth is at present legally entitled to come and stay in this already densely populated country'.

More legislation brought more restrictions but they did not do enough to placate the fears of Enoch Powell and in perhaps the most passionate and most notorious passage of his speech, he said:

It almost passes belief that at this moment 20 or 30 additional immigrant children are arriving from overseas in Wolverhampton alone every week – and that means 15 or 20 additional families a decade or two hence. Those whom the gods wish to destroy, they first make mad. We must be mad, literally mad, as a nation to be permitting the annual inflow of some 50,000 dependants, who are for the most part the material of the future growth of the immigrant-descended population. It is like watching a nation

258

busily engaged in heaping up its own funeral pyre. So insane are we that we actually permit unmarried persons to immigrate for the purpose of founding a family with spouses and fiancés whom they have never seen.

This sense of Britain, the British and British culture being 'swamped' was a deliberately alarmist notion developed by a string of politicians since Powell, some of them prime ministers and holders of high offices of state. But it did not happen, and no rivers foamed with blood. Race riots certainly occurred periodically, and notably in Notting Hill in 1958 and Toxteth in 1981, and racist kill-ings have, tragically, taken place from time to time, but the explosive tensions foreseen by Powell have not become a dominant theme of modern British history. Overtly racist political parties remain on the parliamentary fringes, occasionally gaining seats in the European Parliament or in local government elections. But they have not, thus far, been represented in the legislatures of Westminster, Edinburgh, Cardiff or Belfast.

Instead, what has been striking is the positive reaction to the scaremongering of Powell and others. In 1976 the Race Relations Act brought the Commission for Racial Equality into being and the sort of discrimination that had sometimes disfigured British society became unlawful. And these encouraging trends have taken place against a background of increasing immigration.

One definition of levels of immigration, but not the only one, looks at the country of birth of residents of Britain and between 1991 and 2011, the percentage of people who were born abroad rose sharply. In 20 years it moved from 7% to 13% in England and Wales, and also increased in Scotland and Northern Ireland though less dramatically. The total number of foreign-born people in Britain was calculated at 4,859,000 in 1991 and 6,908,000 in 2011. The list of countries of origin is very long, but 45% of the total came from only ten. India led the table with 694,000, Poland was second with 579,000 and Pakistan third with 482,000.

Since the accession of eight new member states to the European Union in 2004, the number of Polish people living and working in Britain has increased fivefold in only seven years. Polish is now the third most spoken language after English and Welsh. But the nature of this immigration and those people who travel from the European Union member states to find work is different. Many settle, but many also return home after a time, often having saved money so that they can better their or their family's circumstances at home. The total number of what the census calls EU8 (the 8 central and eastern European states who joined the EU in 2004) nationals had risen to 1,038,000 in 2011, while the number from the entire European Union, the EU26, doubled between 2001 and 2011 from 1,094,000 to 2,283,000. Again, many of these people were not permanent residents.

This influx of Europeans is very eye-catching, but difficult to detect on any high street. Presumably their arrival would not have dismayed Enoch Powell or his despondent constituent as much as the black people they saw on the streets of Wolverhampton. Different skin colour allied to a different culture often excites the most visceral of prejudices, but there are fascinating indications that such crude distinctions may be breaking down.

The 2011 census counted a population of over 1 million British Afro-Caribbeans. Recent research has shown that about 50% of men in this community have partners from a different ethnic background, and about a third of British Afro-Caribbean women have made the same choice. There exists speculation that this could become the first ethnic group to 'disappear'. Statistics for the highly mobile immigrant group (to say nothing of those who are here illegally) from the EU8 countries are elusive, but calculations based on the rate of change in the largest ethnic group, those who define themselves as White British, are instructive.

In the 2011 Census this group accounted for 86% or 48.2 million people in England and Wales. In 1991 this percentage was much higher at 94.1%. It dropped to 91.3% in 2001. The depletion rate between 1991 and 2001 was 0.28% per year and between 2001 and 2011, it accelerated to 0.53%. If the rate increases exponentially between 2011 and 2021 to 1% a year, the White British ethnic group could drop to 76% in a decade, and even lower as time goes on. This is of course speculative, but if the trends leading towards a multi-cultural and mixed race society continue, the collective face of Britain could change dramatically and rapidly.

In 1993 the Prime Minister, John Major, made a speech to the

Conservative Group for Europe. He closed with these memorable, nostalgic images:

> Fifty years on from now, Britain will still be the country of long shadows on cricket grounds, warm beer, invincible green suburbs, dog lovers and pools fillers, and, as George Orwell said, 'old maids cycling to Holy Communion through the morning mist', and, if we get our way, Shakespeare will still be read even in school. Britain will survive unamendable in all essentials.

However evocative and comforting, Major's backward glance missed a clear historical truth. Britain has constantly been amended, changed and developed, and one of the principal agencies of that change has been the fluctuating flow of immigration. Over millennia we have been amended by new people and they have turned our history in unpredicted directions. Britain has been the destination for millions of genetic journeys and this story of who we are and where we came from cannot be frozen in time as old maids cycle through the morning mist. It is instead dynamic, an unfinished story without a neat or tidy conclusion. The British will keep arriving. For this long narrative shows something unarguable. We are all the descendants of immigrants, people who brought their DNA, their culture and their future to the farthest north-west edge of Europe.

Bibliography

This is a general reading list for those who may wish to pursue their own course of study. It omits the many learned academic articles available to more specialised interests and also the vast resource that is the Internet.

Bahn, Paul G., *Cave Art*, Frances Lincoln, 2007
Barton, Nick, *Ice Age Britain*, Batsford, 1997
Burley, Robbins, *The Talking Ape*, OUP, 2005
Clarkson, Tim, *The Men of the North*, John Donald, 2010
Cunliffe, Barry (ed.), *The Oxford Illustrated History of Prehistoric Europe*, OUP, 1994
Cunliffe, Barry, *The Ancient Celts*, Penguin, 1997
Cunliffe, Barry (ed.), *The Penguin Atlas of British and Irish History*, Penguin, 2001
Cunliffe, Barry ed., *The Penguin Atlas of British and Irish History*, Penguin, 2003
Cunliffe, Barry, *Europe Between the Oceans*, Yale UP, 2008
Cunliffe, Barry, *Britain Begins*, OUP, 2013
Curtis, Gregory, *The Cave Painters*, Anchor Books, 2006
Davies, Norman, *Europe, A History*, Pimlico, 1997
Davies, Norman, *The Isles*, Papermac, 2000
Dawkins, Richard, *The Ancestor's Tale*, Orion, 2004
Deutscher, Guy, *The Unfolding of Language*, Arrow Books, 2006
Devine, Tom, *The Scottish Nation*, Penguin, 1999
Diamond, Jared, *Guns, Germs and Steel*, Norton, 1997
Fagan, Brian, *The Long Summer*, Granta, 2004

Fagan, Brian, *Beyond the Blue Horizon*, Bloomsbury, 2012

Gaffney, Vince, *Europe's Lost World*, CBA, 2009

Herodotus, *The Histories*, Penguin Classics, 1994

Jordan, Paul, *North Sea Saga*, Longman, 2004

Kennedy, Maeve, *Archaeology*, Hamlyn, 1998

Lewis-Williams, David and Pearce, David, *Inside the Neolithic Mind*, Thames & Hudson, 2008

Lynch, Michael, *Scotland: A New History*, Pimlico, 1991

MacLeod, Mona, *Leaving Scotland*, National Museum of Scotland, 1996

McKie, Robin, *The Face of Britain*, Simon & Schuster, 2006

Miles, David, *The Tribes of Britain*, Weidenfeld & Nicolson, 2005

Miles, David, *The Tribes of Britain*, Phoenix, 2006

Mithen, Steven, *After the Ice*, Weidenfeld & Nicolson, 2003

Morris, Desmond, *Horsewatching*, Ebury Press, 2000

Nicolaisen, W.F.H., *Scottish Placenames*, Batsford, 1976

Oppenheimer, Stephen, *The Origins of the British*, Robinson, 2006

Ostler, N., *Empires of the Word*, HarperCollins, 2005

The Oxford Companion to the Earth, OUP, 2000

Pitts, Mike, *Hengeworld*, Arrow Books, 2001

Ralston, Ian (ed.), *The Archaeology of Britain*, Routledge, 1999

Renfrew, Colin, *Prehistory*, Weidenfeld & Nicolson, 2007

Smyth, A.P., *Warlords and Holy Men*, EUP, 1984

Stringer, Chris, *Homo Britannicus*, Penguin, 2006

Watson, Peter, *The Great Divide*, Harper Perennial, 2013

Index